CHEMICAL ENGINEERING METHODS AND TECHNOLOGY

MOLECULAR DYNAMICS

THEORY, KINETICS AND IMPLEMENTATION

Chemical Engineering Methods and Technology

Additional books in this series can be found on Nova's website
under the Series tab.

Additional e-books in this series can be found on Nova's website
under the e-books tab.

Computer Science, Technology and Applications

Additional books in this series can be found on Nova's website
under the Series tab.

Additional e-books in this series can be found on Nova's website
under the e-books tab.

MOLECULAR DYNAMICS

THEORY, KINETICS AND IMPLEMENTATION

DANIEL E. GARCIA

AND

PAIGE J. GREEN

EDITORS

Nova Science Publishers, Inc.

New York

NOTICE TO THE READER

LIBRARY OF CONGRESS CATALOGING-IN-PUBLICATION DATA

Molecular dynamics : theory, kinetics, and implementation / [edited by] Daniel E. Garcia and Paige J. Green.
    p. cm.
 Includes bibliographical references and index.
 ISBN 978-1-62081-545-8 (soft cover)
 1. Molecular dynamics. I. Garcia, Daniel E. II. Green, Paige J.
 QP517.M65M657 2011
 616.07'9--dc23
                    2012010686

Published by Nova Science Publishers, Inc. † New York

CONTENTS

PREFACE

Molecular dynamics (MD) is a computer simulation of physical movements of atoms and molecules. The atoms and molecules are allowed to interact for a period of time, giving a view of the motion of the atoms. This book presents current research on the theory, kinetics and implementation of molecular dynamics. Topics discussed in this compilation include the molecular dynamics of proteins; molecular dynamics simulations on the extraction of fluid transport properties at the nanoscale; investigation of structural properties of drug-metabolizing enzymes using molecular dynamics simulation; double-pulse laser control of ultrafast optical Kerr effect in liquid; ZnO nano-structures for biosensing; and molecular dynamics simulations of liquid and ionic solvation of carbon tetrachloride.

Chapter 1 - Conformational changes of biomolecules have been proposed as a cause and effect of biological events. Particularly, the content of a specific conformation of a protein in an intracellular localized space has suggested signals addressed to execute specific tasks. Sampling of these structural options, that make of proteins informational molecules, finds in molecular dynamics (MD) a theoretical and computational tool to explain and hypo-thesize the functionality at atomistic level. Conceptually, consider the "final" folded structure as the most functional conformation, possibly finding some obstacle to study the pleiotropicity of proteins and its structural basis. In this order of ideas, the evidence provided by MD in relation to folding and unfolding, where thermodynamic states and kinetic properties characterize conformational changes, sheds light on the pathways towards functionality of a protein. In this chapter, force fields and atomistic inter-actions considered as relevant parts of a MD simulations are reviewed and a typical protocol of stability is explained. Finally, statistical mechanics as a theoretical frame to explore the conformational space is discussed in the context of available computational methods devoted to discover functional domains in proteins.

Chapter 2 - Pharmacokinetic studies play an important role in drug design and development. Recently, many drug candidates have failed to be released into the market owing to pharmacokinetic problems. Thus, drug metabolism should be investigated in the early stages of drug design. Because a large number of drug candidates are screened in the early stages of drug design and development, fast and low-cost estimations of drug metabolism using computational approaches are useful. Structure-based drug design (SBDD), in which three-dimensional (3D) structures of proteins are used for molecular design, is a widely used computational approach. SBDD approaches can be used for drug target as well as for drug-metabolizing enzymes in drug design testing. For example, undesirable inhibition of drug metabolism can be identified and the effects of genetic polymorphism can be predicted for drug candidates using the SBDD approach for drug-metabolizing enzymes. Although 3D structures of proteins are indispensable to the SBDD approach, experimental structures of many drug-metabolizing enzymes are still unknown. Thus, computational procedures for prediction and refinement of the 3D structures of drug-metabolizing enzymes are important. In this chapter, structural refinements of UDP-glucuronyltransferase using molecular dynamics simulations are described. Because there are polymorphisms for this enzyme, its detailed structure is essential for SBDD. Thus, molecular dynamics simulations are expected to be useful for structural refinements of drug-metabolizing enzymes.

Chapter 3 - The optical control of the molecular motions in the neat liquids of carbon tetrachloride CCl_4 and chloroform $CHCl_3$ at room temperature through the non-resonant excitation was enhanced by means of the double-pulse pump-probe technique. When the separation time of the pump pulses and their relative intensity were varied, the amplification or the cancellation of the coherent vibrations of the molecules was achieved. The molecular responses were detected by the time-resolved optically heterodyne-detected optical-Kerr-effect (OKE) technique. It was shown that the time-resolved polarization selective spectroscopy of the molecular vibrational and rotational dynamics in liquid can be based on the double-pulse laser control.

Chapter 4 - ZnO nanostructure is a material that is central for many nanotechnology applications, such as chemical and biological sensors. A systematic molecular dynamics study for the behavior of water droplet and electrolyte solutions interacting with ZnO were done. The contact angle of a water droplet on ZnO polar slabs and nanorods/-tubes array changes significantly as a function of the ZnO-water interaction energy and nano structure geometry. The water contact angle served as a criterion to tune the intermolecular interactions. To recover a hydrophilic surface, voltage range from 1 to 20 volt were applied along the z-axis of the system to simulate the electrowetting behavior case. ZnO nanotube was used to study the permeation of water for equilibrium and applied voltage cases, illustrating the influence of the surface topography and the intermolecular parameters and surface charges on permeation kinetics. The author also studied the ionic currents through ZnO nanotubes simulating the case as field effect transistor (FET) of NaCl, KCl, CaCl2, and MgCl2 electrolyte solution for different concentrations (0.1, 0.5, 1.0, 5.0, and 10.0M) and calculate the solution conductance of these ions by applying a voltage difference (1-20V) along the z-axis of the system. Its possible to achieved, by these molecular dynamics simulations, the characteristic behaviors of ZnO nanostructure-electrolyte solution interactions and how it is suitable to use as ion selective sensor for intracellular micro environment.

Chapter 5 - A flexible body and nine-site fluctuating charge model to describe the intramolecular and intermolecular interactions for carbon tetrachloride is constructed based on the combination of atom-bond electro negativity equalization method and molecular mechanics (ABEEM/MM). The potential parameters are refined to accurately describe the structures and binding energies of $(CCl_4)_2$ and $Na^+(CCl_4)_n$ (n=1-4) clusters, then carried out liquid CCl_4 and Na^+solvationin liquid CCl_4 simulations to examine the potential parameters. The computed density of liquid CCl_4 is in excellent agreement with experimental value. The structures of liquid CCl_4 and Na^+ in liquid CCl_4 can be analyzed by examining the radial distribution functions and angular distribution functions. It is found that the liquid CCl_4 forms an interlocking structure and that a local orientation correlation is observed between neighboring CCl_4 molecules. In the study of Na^+ solvation in liquid CCl_4, It is possible to observe a well-defined solvation shell around Na^+ with five coordinated CCl_4 molecules. It is also found that Na^+ induces a strong local orientation order in liquid CCl_4. Moreover, the calculated diffusion constant of CCl_4 is in reasonable agreement with the experimental result. This work demonstrates that the new potential model is good enough to reproduce properties of liquid $CCl_4$and Na^+ in bulk liquid CCl_4.

In: Molecular Dynamics
Editors: D. E. Garcia and P. J. Green

ISBN: 978-1-62081-545-8
© 2012 Nova Science Publishers, Inc.

Chapter 1

MOLECULAR DYNAMICS OF PROTEINS: TOWARDS FUNCTION IDENTIFICATION VIA STABILITY

Ian Ilizaliturri-Flores,[2] Jorge L. Rosas-Trigueros,[2,4,]* Beatriz Zamora-López,[3] José Correa-Basurto[5], Claudia G. Benítez-Cardoza[1,2] and Absalom Zamorano1,2,†*

1 Programa Institucional De Biomedicina Molecular, Enmh-Ipn, México D.F., México
2 Red De Biotecnología Del Ipn, México D.F., México
3 Departamento De Psiquiatría Y Salud Mental, Facultad De Medicina, Unam, México D.F., México
4 Grupo Interdisciplinario De Inteligencia Artificial Aplicada Al Plegamiento De Proteínas, Escom-Ipn , México D.F., México
5 Laboratorio De Modelado Molecular, Esm-Ipn, México D.F., México

ABSTRACT

Conformational changes of biomolecules have been proposed as a cause and effect of biological events. Particularly, the content of a

---

* Ilzaliturri-Flores and Rosas-Trigueros contributed equally to this chapter

specific conformation of a protein in an intracellular localized space has suggested signals addressed to execute specific tasks. Sampling of these structural options, that make of proteins informational molecules, finds in molecular dynamics (MD) a theoretical and computational tool to explain and hypothesize the functionality at atomistic level. Conceptually, if we consider the "final" folded structure as the most functional conformation, possibly we find some obstacle to study the pleiotropicity of proteins and its structural basis. In this order of ideas, the evidence provided by MD in relation to folding and unfolding, where thermodynamic states and kinetic properties characterize conformational changes, sheds light on the pathways towards functionality of a protein. In this chapter, force fields and atomistic interactions considered as relevant parts of a MD simulations are reviewed and a typical protocol of stability is explained. Finally, statistical mechanics as a theoretical frame to explore the conformational space is discussed in the context of available computational methods devoted to discover functional domains in proteins.

1. INTRODUCTION

Proteins perform important tasks including direct their own folding. Exposition to variety of stressful conditions may promote proteins misfolding, as is shown in tumor development wherein mutations interfere with the appropriate role of tumor-suppressor proteins and oncogenes. Alterations of catalytic efficiency, loos of binding sites or of functional form, are included as consequences. Some examples such as Src family kinases, p53, mTOR, and C-terminus of HSC70 interacting protein (CHIPs) are associated with protein misfolding and tumorigenesis. Therefore, a therapeutic strategy could be related to repair or eliminate protein misfolding (Nagaraj et al. 2010). Another interesting instance is the fusion oncoprotein formed by Promyelocytic leukemia (PML) and the retinoic acid receptor (RAR) that is associated to the transformation of acute promyelocytic leukemia (APL). When this protein fusion is misfolded promotes the same in the nuclear receptor corepressor (N-CoR) protein, a corepressor essential for the growth-suppressive function of several tumor-suppressor proteins. This "trans-misfolding" is mediated by inducing an anomalous post-translational modification. (Khan 2010).

Likewise, oxidative stress, characterized by the overproduction of reactive oxygen species (ROS) and reactive nitrogen species (RNS), has been proposed to be generated by misfolded proteins. In turn, a harmful loop arises from production of oxidatively modified proteins, with adverse physiological consequences. (Gregersen & Bross 2010). During aging and in neuro-degenerative diseases components of the protein quality control system, become functionally impaired, and its impact could be greater. Investigation in disabling diseases such as Alzheimer's disease, Parkinson's disease, transmissible spongiform encephalopathies, familial amyloid polyneuropathy, Huntington's disease, and type II diabetes has suggested that protein mis-folding could be cause of cell death and tissue disfunction. In these pathologies, large quantities of incorrectly folded proteins undergo aggre-gation (Uversky 2009; Silva et al. 2010). Delve into how proteins folds is the first approach to explain how they misfold, and in consequence, how to avoid it or restore a functional form (Kelley et al. 2008).

Understanding of the process of folding protein involves the development of an interdisciplinary research: chemical, physical and biological foundations. When interpreted in conjunction with experiment results, computational simulations with atomic resolution provide information of protein dynamic behavior and non-equilibrium events like folding and unfolding.

A widely implemented technique termed Molecular Dynamics (MD) models the motion in molecular systems. By the time-dependent integration of classical equations, novel atomic positions are obtained each discrete time step, assuming that is sufficiently small (2 fs, or less). Since based on a molecular mechanics force field the potential energy and the force are calculated. Construction of functionality hypothesis of domains, from detailed models of specific peptides and proteins is of interest, to identify relevant structures and explore if their thermodynamic parameters could be associated. In this chapter we review some topics in the MD approach and its potential to investigate protein functional domains. These ideas contribute to a more efficient structure prediction and to obtain optimized models for molecular kinetics that could aid in bioinformatical drug and protein design (Bowman & Vijay S Pande 2009; Ozkan et al. 2007).

2. WHY DOES FOLDING/UNFOLDING DRAW OUR ATTENTION?

Proteins are involved in every single biological process of a living system. They are synthesized on ribosomes, as linear chains of usually some hundreds of aminoacid residues, in a specific order dictated by DNA sequence. In order to perform their various functions most proteins must be folded. It means that they have adopted a specific stable, well-packed three-dimensional structure, known as the native conformation (Anfinsen 1973). Commonly the native state corresponds to the most thermodynamically stable structure at physiological conditions (Christopher M Dobson 2003). Nevertheless, the conformational space that a polypepetide would have to sample in order to find the global minimum is enormous and might involve several local metastable structures on or off pathway as well as several kinetic traps. Therefore, it is not possible that a polypeptide would systematically or randomly search within this conformational space to get its native structure because it would take an astronomical lenght of time, and usually protein folding takes seconds to minutes (Christopher M Dobson 2003).

Since the early 1960s Anfinsen's experiments showed that all the information that proteins need to self-assembly is contained within the sequence itself and there is not need of any other molecule to carry out the folding process (Anfinsen 1973). In addition, a nascent polypeptide does not need to explore the whole conformational space during its transition from a random coil conformation to a native structure; instead the energy surface or 'landscape' is funneled. This means, that the energy decreases as a function of nativeness. The same is true for the configurational entropy of the polypeptide that decreases as folding progresses by maximizing hydrogen-bonding of polar groups and by burying hydrophobic residues away from the aqueous environment (Ionov et al. 1990). As a consequence, the nascent polypeptide samples only a small number of conformations within the conformational space during the protein folding process (Wolynes et al. 1995; K A Dill & Chan 1997; Dinner et al. 2000). In proteins the energy landscape is encoded by the amino-acid sequence, whereas landscapes of random heteropolymers does not have these features (Bryngelson et al. 1995). It means that natural selection has pushed proteins to fold rapidly and efficiently (Christopher M Dobson 2003). Although theory and experiment have provided answers to many of the fundamental questions of protein folding, it is not yet clear exactly how the sequence encodes such characteristics. It is very likely that the pattern of hydrophobic and polar residues favours preferential interactions of specific residues as the structure becomes increasingly compact (Lin et al. 2011; Vendruscolo et al. 2001). Once the correct topology has been achieved, in most cases the native conformation will be achieved during the final stages of folding. A still remaining challenge is an accurate, high-resolution picture of folding mechanism (Best 2012). All these ideas encourage our interest to better understand every step involved in the folding process by both experimental and theoretical procedures.

Furthermore, within the cell the folding process of a nascent polypeptide can be affected by several factors beyond amino acid sequence, for example the physicochemical conditions of the microenvironment where this reaction takes place, as well as transient interactions with other proteins and ligands as well as the ribosome. Also, we recently have learned that the translation rate has an important effect on protein folding, and consequently in the structure and function of the protein product. In particular, the use of synonymous codons in the mRNA and the availability of tRNAs can modulate translation kinetics, affecting the folding, the structure and the biological activity of proteins (Marin 2008; Deane & Saunders 2011; Zhang et al. 2009). Moreover, some proteins fold in specific compartments, such as mitochondria or the endoplasmic reticulum (ER), after trafficking and translocation through membranes. Consequently, during the folding process the partially folded conformations of most proteins expose to the solvent hydrophobic regions that in the native structure are buried. These hydrophobic patches might represent sticky surfaces prone to incorrectly interact with other molecules within the crowded environment of a cell. Fortunately, living systems have developed a series of strategies to avoid incorrect interactions between incompletely folded proteins. The most relevant guardians of the folding process are molecular chaperones (Hartl & Hayer-Hartl 2002; Young et al. 2004; Young et al. 2004). Some molecular chaperones have been widely studied, like GroEL and its co-chaperone GroES. It is known that GroEL as well as some other chaperones of this class contain a cavity where the partially folded chain can enter and complete its folding process, hiding the sticky surfaces from the environment (Young et al. 2004). It is clear that molecular chaperones neither contain information about the folding process, nor are capable of accelerating the reaction. Still, there are some other molecules that are able to increase the rate of individual steps of the folding process. The most relevant examples are peptidylprolyl isomerases that catalyze the cis–trans isomerization of peptide bonds involving proline residues and disulphide isomerases which improve the rapidity of formation and reorganization of disulphide bonds (Schiene & Fischer 2000).

However, sometimes protein folding can go wrong and some others proteins are not capable to remain correctly folded resulting in the malfunctioning of living systems and hence to disease (C M Dobson 2001). The lack of function of incorrectly folded proteins has been associated to diseases such as cystic fibrosis and some types of cancer (Thomas et al. 1995; Bullock & Fersht 2001; Suad et al. 2009). Many of these types of disorders arise from mutations in the sequence of the respective protein, and in most cases are heritable. Another scenario arises from misfolded proteins that elude the protective mechanisms and accumulate forming aggregates that deposit inside the cells or in the intracellular space of several tissues. Some examples of such affections are Alzheimer's and Parkinson's diseases, the spongiform encephalopathies and type II diabetes are caused by the accumulation of misfolded proteins in a range of tissues including brain, heart and spleen, among others (C M Dobson 2001; Horwich 2002; Christopher M Dobson 2002; Chatterjee & Van Marck 2006).

Considering all this, there is not doubt that it is essential to keep growing our knowledge and understanding of protein folding and misfolding mechanisms from a multidisciplinary point of view. The day that this tremendous task is achieved it will be possible to predict the native confor-mation of any new sequence based solely on the primary structure. Under-standing in detail the folding mechanism of proteins would also allow us to identify factors that encode and stabilize the characteristics of proteins in order to modify them or design them rationally. Fortunately, theory and experiment are cooperatively helping us to understand this important unsolved problem in molecular biology and we will be able to better use the information becoming available from the mapping of genomic sequences.

3. INTER- AND INTRA-MOLECULAR INTERACTIONS INVOLVED IN THE STABILITY

There is a balance between non-covalent interactions and thermodynamic forces in the structural stability and function of proteins. Typically the interactions which are considered to dominate protein stability are depending of the hydrophobic effects, hydrogen bonds, electrostatic interactions among others located at protein internal surfaces (Abaturov & Varshavskiĭ 1978; Rosas-Trigueros et al. 2011). It is important to mention that these non-bond interactions give entropy properties and consequently those governing the conformational degree as well as the energy of dihedral and bond angle strain (Wang et al. 2011). These inter- and intra-molecular interactions that influence in the protein stability can be evaluated by experimental and computational procedures. These studies help to understand the physical and chemical forces as the key to the prediction of fold protein structure, recognition, function among others biological properties. It is known that proteins exhibit marginal structural stabilities that are the sum of a great number of intermolecular non bonded interactions (Ken A. Dill 1990). Average values for the Gibbs free energy of stabilization of medium size globular proteins are the order of 50 kJmol^{-1} (R Jaenicke 2000). These intermolecular non bonded interactions are listed in the Table 1.

Table 1. Interactions that stabilize proteins

Interaction	Distance	Description	DG (kJ/mole)	Reference
Covalent bond	1.5 Å	involve the sharing of a pair of valence electrons by two atoms	356	(Gagneux 2004)
Salt Bridge	2.8 Å	They are hydrogen bonds which link charged partners	12.5-30	(Jelesarov & Karshikoff 2009)
Hydrogen Bond	3.0 Å	They take place between pairs of atoms only if one of them is a proton donor and the other one is a proton acceptor	2-6	(Jelesarov & Karshikoff 2009)
Long-range electrostatic interaction	-	Depends on dielectric constant of medium	Depends on distance and environment	(Gagneux 2004)
Van der Waals interaction	3.5 Å	They are considered as weak attractive interactions which become appreciable only when the interacting molecules are neutral and non-polar.	4-17	(Gagneux 2004)

The importance of hydrogen bonds in stabilizing regular secondary structure elements and tertiary folds have been study widely (Pauling et al. 1951). The helix is stabilized by hydrogen bonding between amine (hydrogen donator, HD) and carbonyl groups (hydrogen acceptor, HA) of the backbone residues making during each (each 3.4 residues) turnover displacing the side chain outside the axes. Whereas that in the other common secondary structure, beta sheet is stabilized by hydrogen bonds between the amine groups (HD) of one backbone residue with the adjacent parallel or antiparallel carbonyl groups (HA) of backbone. To clear the hydrogen bond interaction, they are mediated between a hydrogen located on heteroatom and one polar atom such as oxygen, nitrogen, sulphur, and also can be with halogen atoms. There is other scarcely studied hydrogen interaction, hydrogen-pi interaction which is mediated between the hydrogen located on heteroatom and aromatic rings. The hydrogen bond interactions are very common in intra- or inter-protein interactions (from backbone), and they can mediate the interaction between protein or DNA with small ligands. However, they are in greater number in the protein surfaces of globular protein that is in contact with water environment, from side chain polar. Intramolecular hydrogen bonds become lost in the protein unfolding and are replaced by new ones with water (Privalov 2009). The contribution of the hydrogen bonds to structural stability is evaluated on the basis of model compounds or by mutagenesis experiments (Schweiker & Makhatadze 2009).

Kauzmann has stressed the fact that almost half of amino acid residues in proteins are hydrophobic (Kauzmann 1959). The association of the non-polar molecules, which is the essence of hydrophobic interactions, is a result of tendency of the system to increase its entropy. Due to the hydrophobic effect, the non-polar side chains avoid contact with water and tend to assemble close to each other. As a result, the polypeptide chain collapses so that the hydrophobic residues form the hydrophobic core of the protein molecule. There are several hydrophobic pockets composed by hydrophobic residues such as Val, Leu, Ile or aromatic residues. Essentially, they are located at intra-protein surfaces and can maintain the globular fold structure. These kinds of residues are very common on transmembrane domains as occur with the known G protein coupled receptors (Kobilka 2007). Also, there are proteins that recognize hydrophobic ligands that need to be metabolized as occur with the cytochrome 450 (Lampe et al. 2010) or chloroperoxidase (Sundaramoorthy et al. 1995) among other examples.

The hydrophobic effect is strongly temperature-dependent, and it is considerably weaker and perhaps even destabilizing at low temperatures than at elevated temperatures (Gross & Rainer Jaenicke 1994).

The important role of electrostatic interactions (salt bridges) in proteins becomes evident at any pH-dependent process: pH-regulation of enzyme activity, acid and alkaline denaturation, protein substrate/inhibitor interactions, etc. These kind of interactions are very common on proteins at pH close to 7 where the carboxyl groups of Glu and Asp residues have negative charge whereas that Lys, Arg, and His have positive charges. The electrostatic interactions are named sometimes as salt bridge and can form a hydrogen bond with polar groups such as OH, SH, NH, which can be located in proteins or ligands. It is unlikely that salt bridges are dominant factors governing protein stability. Nevertheless, proteins from thermophiles and hyperthermophiles exhibit more, and frequently networked, salt bridges than proteins from the mesophilic counterparts. Increasing the thermal (not the thermodynamic) stability of proteins by optimization of charge-charge interactions is a good example for an evolutionary solution utilizing physical factors. In contrast to the other types of non-covalent interactions they are short- and long-range and dependent on ionization equilibrium (pKa), which in turn are governed by electrostatic interactions. These non-covalent interactions also participate in ligand-receptor associations, stabilizing tertiary structure, and switching active and inactive conformations in G protein-coupled receptors (Kobilka 2007).

There are several physical properties that influence importantly in the macromolecules that can modify their activity and folds. The effects of pressure on protein structure and function can vary dramatically depending on the magnitude of this, the reaction mechanism (in the case of enzymes), and the overall balance of forces responsible for maintaining the protein structure. Interactions between the protein and solvent are also critical in determining the response of a protein to pressure. As a potential denaturant of proteins, pressure has different effects depending on the structural characteristics of the biomolecule (Gross & Rainer Jaenicke 1994; Boonyaratanakornkit et al. 2002). On the other hand, hydrostatic pressure has been used to increase the activity and/or stability of several enzymes (Fukuda & Kunugi 1984; Hei & D S Clark 1994; Michels & D S Clark 1997).

Related to aromatic interactions can be considered with the hydrophobic interactions, however, it is required be mentioned due to their interest for protein-protein interactions and sometimes for ligand recognition. In particular, intra- and inter-protein contacts are essentially mediated by Tyr, Phe or Trp residues (Sonavane & Chakrabarti 2008).

4. FORCE FIELDS TO SIMULATE PROTEIN STABILITY AND THEIR IMPLEMENTATIONS

The combination of a potential energy function with the parameters used in that function to describe the relationship of the structure to the energy yields a force field. The quality of the force field and its implementation are of the utmost importance for the ability of MD simulations to reproduce experimentally accessible properties. The most commonly used force fields for MD simulations of proteins are OPLS-AA (Jorgensen & Tirado-Rives 1988), CHARMM (MacKerell, et al. 1998) and AMBER (Cornell et al. 1995), which were developed with particular emphasis on the treatment of proteins

(Mackerell 2004) and have yielded similar results in simulations of peptidic chains with 165 amino acids or less (Price & Charles L Brooks 3rd 2002).

The AMBER force field was among the earliest parameterization projects where polar hydrogens were explicitly represented (Ponder & Case 2003). The partial charges were derived from *ab initio* calculations using a restrained electrostatic potential model and most other parameters were taken from crystal structures while only a few were fitted to experimental information (Cornell et al. 1995). Recent developments have focused on the improvement of the torsion potentials (Lindorff-Larsen et al. 2010).

The OPLS-AA force field was developed from the start with the idea of using it for MD simulations of biomolecules (Jorgensen & Tirado-Rives 1988) and was later improved for the same purpose (Kaminski et al. 2001). OPLS-AA is based on the AMBER force field (Weiner et al. 1984), and focuses on the intramolecular nonbonded interactions and replaces the corresponding terms in AMBER's potential energy function. The partial atomic charges were based on supramolecular data from Hartree-Fock *ab initio* calculations. Many other parameters were recalculated with the aim of reproducing the experimental densities and heats of vaporization for several organic liquids (Jorgensen & Tirado-Rives 1988).

The development of the CHARMM force field has not been particularly focused on the simulation of proteins, but at handling molecular systems of many different classes, sizes, and levels of heterogeneity and complexity (B R Brooks et al. 2009). Partial atomic charges were determined in a similar fashion as for OPLS-AA and the Lennard-Jones parameters were refined to reproduce densities and heats of vaporization of liquids as well as unit cell parameters and heats of sublimation for crystals (Ponder & Case 2003).

Certain differences have been identified during the performance of these force fields. For instance, the energetics from stable conformations of peptides in gas phase found using OPLS-AA and CHARMM compare well with high-level *ab initio* calculations (Beachy et al. 1997). Nevertheless, OPLS-AA outperformed both AMBER and CHARMM in free energy of hydration calculations (Michael R. Shirts et al. 2003).

Some concern has been raised about a possible bias in these force fields. CHARMM with the CMAP correction has been found to show a strong alpha-helical bias. Also, some versions of AMBER have been found to have an excessive propensity towards visiting alpha-helical conformations (Best et al. 2008). However, simulations using AMBER99SB have successfully predicted the number of structurally major clusters, their relative and absolute stabilities and even the experimentally found secondary structure of the backbone which is remarkable since it is a 3-10 helix, thus distinct from the most commonly found alpha-helix, beta-sheet and coil. On the other hand, a beta-structure bias has been suggested for OPLS-AA and CHARMM22 (Patapati & Glykos 2011; MacKerell, et al. 1998).

Protein MD simulations are usually performed with explicit water molecules added to the system since specific water-protein interactions are often important (Mackerell 2004). The single point charge model with energy correction (SPC/E) (Berendsen et al. 1987) considers three interaction sites placed at fixed distances from each other and models water geometry as an ideal tetrahedron. An important family of water models originated with the TIP3P model, which also has three interaction sites and models water geometry after the experimentally observed intermolecular distances and angles (Jorgensen et al. 1983). Later developments on this model added interaction sites to obtain the TIP4P, TIP5P and TIP6P models (Mahoney & Jorgensen 2000; Nada & van der Eerden 2003). While all these models yield reasonable agreement for bulk water at ambient temperature, the agreement of the SPC/E model with the experiment in surface tension calculations was only paralleled by the six-site TIP6P model (Chen & Smith 2007).

5. MD SIMULATIONS OF PROTEINS IN PRACTICE

Classical molecular dynamics (MD) simulation is certainly an approximation of a physical system such as macromolecules. From an experimentalist point of view, it is fiction trying to emulate reality. However, it must be noted that carefully performed MD simulations can add significant hypothesis in biomolecules of scientific interest. Despite all the inherent approximations, MD simulations can perform very well when certain assumptions are met. In general, MD solves the Newton's equation of motion of a system, allowing to study protein folding/unfolding, protein stability, conformation structural, molecular recognition, etc. Commonly, used software packages in biomolecular simulation, as we mentioned, are AMBER (Case et al. 2005), CHARMM (Bernard R. Brooks et al. 1983), NAMD (Phillips et al. 2005), and GROMACS (Van Der Spoel et al. 2005). GROMACS and NAMD have gained considerable popularity during latest years, mainly because they are freely available for academic users and their vastly improved computational performance. These simulation packages come along with analysis tools, which can be very time consuming sometimes. There is also a continuous interest in developing MD analysis tools to facilitate certain post-simulation tasks. It is worth noting that storing and managing MD results is not a trivial task. Full solvated MD trajectories can occupy several gigabytes of disk space, and a considerable amount of physical memory is needed to perform analysis tasks, even though current evolution in desktop computers power has contributed to the application of MD simulations.

A MD protocol usually involves setting several thermodynamic state variables of the system to a fixed value while allowing other system properties to fluctuate. For instance, a simulation under NPT conditions is achieved by defining a constant number of atoms (N), a constant pressure (P), and a fixed temperature (T). In this case, the volume of the system fluctuates to maintain the desired conditions.

The first step in the analysis of the results from an MD simulation is to analyze the general properties of the system such as the total energy and the box-volume. These measurements should asymptotically converge to a constant value for the system to reach stability. Afterwards, several structural analyses can be performed to gain insight into the behavior of the system.

The root mean square deviation (RMSD) of the atoms in a molecule with respect to a reference structure is calculated by least-square fitting the structures from the MD simulation to the reference structure and calculating the square root of the average of the squared distances of corresponding atoms weighted by the mass of the atoms. Both the fitting and the calculation are often performed using only the alpha-carbons of each amino acid and hence RMSD becomes a measure of backbone conformational movement. RMSD values should converge to a stable value; however, abrupt changes in their behavior could point to conformational changes in a peptidic chain.

Another common analysis is the radius of gyration, which is inversely related to the compactness of the structure. Additionally, the evolution of the solvent accessible surface area can reveal the exposure of originally hidden amino acids. The solvation of certain domains of a protein can be strongly relevant in the investigation of its structural stability. Furthermore, changes in the root mean square fluctuation of atomic positions during the MD simulation can help to identify the most stable and the most fragile points in the protein scaffold.

Other analyses that can be performed with the aforementioned software packages include the estimation of NMR chemical shifts, radial distribution and pair correlation between particles, self-diffusion coefficient of particles and order parameters to investigate the orientation of long carbon chains. For the case of MD simulations of proteins, the evolution of secondary structure elements and the Ramachandran plot can also contribute to the understanding of the intramolecular interactions.

6. SIMULATING FOLDING

To date the dynamics of the native state and folding/unfolding pathways of thousands of proteins from hundreds of organisms have been analyzed. These studies cover the majority of known folds, and a wide range of functions. There have been several efforts to gather this information (van der Kamp et al. 2010). All these studies have help us to better understand the function, interactions with other proteins or ligands as well a structural features that are important in the stability of several proteins.

Table 2. Examples of proteins whose stability has been studied by MD

Protein	Organism	PDB code	Structural classification	Reference
Proteins Mainly Stabilized by Hydrogen Bonds				
phospholipase A2	Streptomyces violaceoruber	1FAZ	All α	(Jain & Sankararamakrishnan 2011)
alginate lyase	Corynebacterium sp.	1UAI	All α	(Jain & Sankararamakrishnan 2011)
antibacterial protein	Streptomyces carzinostaticus	1NOA	All β	(Jain & Sankararamakrishnan 2011)
Proteins Mainly Stabilized by Salt Bridges				
Bcl-2-associated X (Bax)	Homo sapiens	1F16	All α	(Rosas-Trigueros et al. 2011)
Ribosomal Protein L9 (NTL9)	Geobacillus stearothermophilus	2HBB	$(\alpha + \beta)$	(Shen 2010)
subtilisin-like protease	*Vibrio* sp	1SH7	$(\alpha + \beta)$	(Tiberti & Papaleo 2011)
thermophilic thermitase	*Thermoactinomyces vulgaris*	1THM	(α / β)	(Tiberti & Papaleo 2011)
Proteinase K	*Tritirachium album*	1IC6	(α / β)	(Tiberti & Papaleo 2011)
Proteins Mainly Stabilized by Hydrophobic Effect				
Villin headpiece domain	*Gallus gallus*	1VII	All α	(Petrov & Zagrovic 2011)
Cold shock protein	*Bacillus caldolyticus*	1C90	(α / β)	(Stumpe & Grubmüller 2009)

In the Table 2 we present some reports of proteins which stability and folding/unfolding mechanism has been analyzed by Molecular Dynamics Simulation. We focused in showing some examples of non-covalent forces that are the main responsible for the stability of certain proteins.

During 1980s there was a motivation to study the formation of secondary structural elements since their participation had been considered important in determining protein folding pathways. Various of these studies focused on isolated peptides in solution to evaluate their tendency to form turns, helices, and beta-hairpins (Dyson, Sayre, et al. 1992; Dyson, Merutka, et al. 1992). Later, the topology was investigated if it is a dominant source of heterogeneity in the transition states. And not only evidence of some transition state ensemble is determined by the final form was shown, but also for larger proteins that are not two-state folded, some intermediates are topological dependent (Plaxco et al. 1998). Furthermore, this appraisal appeared to agree with the correlation between contacting residues in the native structure and folding rates for single domain proteins (Chan 1998). These and other informative parameters as well as their correlation remain to be investigated.

7. Statistical Mechanics as a Theoretical Framework

Molecular Dynamics simulations have a sound basis in statistical mechanics and classical physics. Under these theories folding pathways are studied in the frame of thermodynamic stability of the partially folded intermediates, considering the atomic motions through which the protein structures are sampled (McCammon et al. 1977).

Because the experimental observation of the motions of protein atoms at nanosecond timescale is, up to now, impossible, the observable measured in the experiment can only ever be a time-averaged quantity. Likewise, by the sampling of the conformational space, a molecular mechanics potential permits to gather geometric and thermodynamics parameters that will be ergodic average quantities, *i. e.*, a quantity whose time average converges to its respective value in the space of configurations (Cooke & Schmidler 2008). The entropy S and the Helmholtz free energy A are fundamental quantities in statistical mechanics and in structural biology. Entropy gives information about the energy that cannot be used and, therefore, could not participate as a driving force in protein folding. In turn, free energy is a state function that measures the spontaneity of the reaction interpreted as stability. However the calculation of S and A from molecular dynamics trajectories is not a trivial problem (Meirovitch 2007). Also, folding rate and stability proteins are influenced by hydrogen bond formation, solvation, hydrophobic core formations, and ion pairs. For this reason, the configurational energy and the supplied by the environment will permit achieve and maintain these interactions, and to trigger or not, important structural changes.

To overcome the limitations of insufficient sampling, the development and implementation of algorithms have been pursued. Parallel replicas of a single simulation, with a statistical treatment has been proposed as straightforward and generalizable method (M R Shirts & V S Pande 2001). Likewise, another strategy has involved parallel sampling but with distinct initial conditions; studies of relaxation behavior, transition rate and equilibrium have also applied this method (Philippe Ferrara et al. 2000; P Ferrara & A Caflisch 2000; Cavalli et al. 2002). In another reports, independent simulations are started from the same conformational basin and when one of these simulations exits a basin, all the other simulations are restarted from the new basin position (Voter 1998; M R Shirts & V S Pande 2001). In a recurrent method called umbrella sampling, one or more independent simulations are performed with modified potentials, giving rise to a phase space that must be corrected to evaluate the differential results due to the potential modification (Charles L Brooks 3rd 2002; Shea & C L Brooks 3rd 2001). This sampling challenge, also has combined weighted histogram techniques to asses thermodynamic properties via a clustering of conformations by a folding/unfolding protocol (Charles L Brooks 3rd 2002).

Interestingly, there is a discussion about the paradigm of the structure-function relationship, namely, this is only an idealization because disorder itself is an important driving force through the second law of thermodynamics, however, the functional role reported of disordered proteins is an issue that must be addressed (Jacobs 2010). The rapid evolution towards equilibrium of the simple and isolated systems contrasts with resistance of living systems to reach equilibrium, even sometimes increase their order. Although, this does not contradict the second law of thermodynamics, since they are open systems. Therefore, the description of the pathways and mechanisms in which the stability of the proteins participates, increasing the entropy of the universe and constructing ordered subsystems, is an open problem (Lazaridis & Martin Karplus 2003).

CONCLUSION

Computation analysis based on a molecular interpretation of folding free energy landscape provides quantitative pictures of the folded and unfolded states. Also, the specific role of the water in the late stages of folding permits hypothesize about local dynamic where entropic forces, along with another enthalpic strength, give rise to new atomic configuration possibly functional. In bridging the gap between computational and experimental results, the calibration of the force fields play an important role to improve simulations methods. To this extent, we will hone our understanding of folding protein from microscopic explanations.

REFERENCES

Abaturov, L.V. & Varshavskiĭ, I.M., 1978. [Equilibrium dynamics of the 3-dimensional structure of globular proteins]. *Molekuliarnaia Biologiia*, 12(1), pp.36-46.

Anfinsen, C.B., 1973. Principles that govern the folding of protein chains. *Science (New York, N.Y.)*, 181(4096), pp.223-230.

Beachy, M.D. et al., 1997. Accurate ab Initio Quantum Chemical Determination of the Relative Energetics of Peptide Conformations and Assessment of Empirical Force Fields. *Journal of the American Chemical Society*, 119(25), pp.5908-5920.

Berendsen, H.J.C., Grigera, J.R. & Straatsma, T.P., 1987. The missing term in effective pair potentials. *The Journal of Physical Chemistry*, 91(24), pp.6269-6271.

Best, R.B., 2012. Atomistic molecular simulations of protein folding. *Current Opinion in Structural Biology*. Available at:
http://www.ncbi.nlm.nih.gov/pubmed/22257762
[Accedido febrero 3, 2012].

Best, R.B., Buchete, N.-V. & Hummer, G., 2008. Are current molecular dynamics force fields too helical? *Biophysical Journal*, 95(1), pp.L07-09.

Boonyaratanakornkit, B.B., Park, C.B. & Clark, Douglas S, 2002. Pressure effects on intra- and intermolecular interactions within proteins. *Biochimica Et Biophysica Acta*, 1595(1-2), pp.235-249.

Bowman, G.R. & Pande, Vijay S, 2009. The roles of entropy and kinetics in structure prediction. *PloS One*, 4(6), p.e5840.

Brooks, B R et al., 2009. CHARMM: the biomolecular simulation program. *Journal of Computational Chemistry*, 30(10), pp.1545-1614.

Brooks, Bernard R. et al., 1983. CHARMM: A program for macromolecular energy, minimization, and dynamics calculations. *Journal of Computational Chemistry*, 4(2), pp.187-217.

Brooks, Charles L, 3rd, 2002. Protein and peptide folding explored with molecular simulations. *Accounts of Chemical Research*, 35(6), pp.447-454.

Bryngelson, J.D. et al., 1995. Funnels, pathways, and the energy landscape of protein folding: a synthesis. *Proteins*, 21(3), pp.167-195.

Bullock, A.N. & Fersht, A.R., 2001. Rescuing the function of mutant p53. *Nature Reviews. Cancer*, 1(1), pp.68-76.

Case, D.A. et al., 2005. The Amber biomolecular simulation programs. *Journal of Computational Chemistry*, 26(16), pp.1668-1688.

Cavalli, A., Ferrara, Philippe & Caflisch, Amedeo, 2002. Weak temperature dependence of the free energy surface and folding pathways of structured peptides. *Proteins*, 47(3), pp.305-314.

Chan, H.S., 1998. Protein folding. Matching speed and locality. *Nature*, 392(6678), pp.761-763.

Chatterjee, S. & Van Marck, E., 2006. Human prion disease hypothesis does not justify the origin of bovine spongiform encephalopathy. *Journal of Postgraduate Medicine*, 52(3), pp.223-225.

Chen, F. & Smith, P.E., 2007. Simulated surface tensions of common water models. *The Journal of Chemical Physics*, 126(22), p.221101.

Cooke, B. & Schmidler, S.C., 2008. Statistical prediction and molecular dynamics simulation. *Biophysical Journal*, 95(10), pp.4497-4511.

Cornell, W.D. et al., 1995. A Second Generation Force Field for the Simulation of Proteins, Nucleic Acids, and Organic Molecules. *Journal of the American Chemical Society*, 117(19), pp.5179-5197.

Deane, C.M. & Saunders, R., 2011. The imprint of codons on protein structure. *Biotechnology Journal*, 6(6), pp.641-649.

Dill, K A & Chan, H.S., 1997. From Levinthal to pathways to funnels. *Nature Structural Biology*, 4(1), pp.10-19.

Dill, Ken A., 1990. Dominant forces in protein folding. *Biochemistry*, 29(31), pp.7133-7155.

Dinner, A.R. et al., 2000. Understanding protein folding via free-energy surfaces from theory and experiment. *Trends in Biochemical Sciences*, 25(7), pp.331-339.

Dobson, C M, 2001. The structural basis of protein folding and its links with human disease. *Philosophical Transactions of the Royal Society of London. Series B, Biological Sciences*, 356(1406), pp.133-145.

Dobson, Christopher M, 2003. Protein folding and misfolding. *Nature*, 426(6968), pp.884-890.

Dobson, Christopher M, 2002. Getting out of shape. *Nature*, 418(6899), pp.729-730.

Dyson, H.J., Merutka, G., et al., 1992. Folding of peptide fragments comprising the complete sequence of proteins. Models for initiation of protein folding. I. Myohemerythrin. *Journal of Molecular Biology*, 226(3), pp.795-817.

Dyson, H.J., Sayre, J.R., et al., 1992. Folding of peptide fragments comprising the complete sequence of proteins. Models for initiation of protein folding. II. Plastocyanin. *Journal of Molecular Biology*, 226(3), pp.819-835.

Ferrara, P & Caflisch, A, 2000. Folding simulations of a three-stranded antiparallel beta -sheet peptide. *Proceedings of the National Academy of Sciences of the United States of America*, 97(20), pp.10780-10785.

Ferrara, Philippe, Apostolakis, J. & Caflisch, Amedeo, 2000. Thermodynamics and Kinetics of Folding of Two Model Peptides Investigated by Molecular Dynamics Simulations. *The Journal of Physical Chemistry B*, 104(20), pp.5000-5010.

Fukuda, M. & Kunugi, S., 1984. Pressure dependence of thermolysin catalysis. *European Journal of Biochemistry*, 142(3), pp.565-570.

Gagneux, P., 2004. Protein Structure and Function. *Journal of Heredity*, 95(3), pp.274-274.

Gregersen, N. & Bross, P., 2010. Protein misfolding and cellular stress: an overview. *Methods in Molecular Biology (Clifton, N.J.)*, 648, pp.3-23.

Gross, M. & Jaenicke, Rainer, 1994. Proteins under pressure. The influence of high hydrostatic pressure on structure, function and assembly of proteins and protein complexes. *European Journal of Biochemistry*, 221(2), pp.617-630.

Hartl, F.U. & Hayer-Hartl, M., 2002. Molecular chaperones in the cytosol: from nascent chain to folded protein. *Science (New York, N.Y.)*, 295(5561), pp.1852-1858.

Hei, D.J. & Clark, D S, 1994. Pressure stabilization of proteins from extreme thermophiles. *Applied and Environmental Microbiology*, 60(3), pp.932-939.

Horwich, A., 2002. Protein aggregation in disease: a role for folding intermediates forming specific multimeric interactions. *The Journal of Clinical Investigation*, 110(9), pp.1221-1232.

Ionov, A.I., Fisher, A.A. & Shubich, M.G., 1990. [The diagnostic information value of determining the cytochemical properties of the neutrophils from the blood and synovial fluid of patients with rheumatoid arthritis and osteoarthrosis deformans]. *Terapevticheskiĭ Arkhiv*, 62(5), pp.51-55.

Jacobs, D.J., 2010. Ensemble-based methods for describing protein dynamics. *Current Opinion in Pharmacology*, 10(6), pp.760-769.

Jaenicke, R, 2000. Stability and stabilization of globular proteins in solution. *Journal of Biotechnology*, 79(3), pp.193-203.

Jain, A. & Sankararamakrishnan, R., 2011. Dynamics of noncovalent interactions in all-α and all-β class proteins: implications for the stability of amyloid aggregates. *Journal of Chemical Information and Modeling*, 51(12), pp.3208-3216.

Jelesarov, I. & Karshikoff, A., 2009. Defining the role of salt bridges in protein stability. *Methods in Molecular Biology (Clifton, N.J.)*, 490, pp.227-260.

Jorgensen, W.L. et al., 1983. Comparison of simple potential functions for simulating liquid water. *The Journal of Chemical Physics*, 79(2), p.926.

Jorgensen, W.L. & Tirado-Rives, J., 1988. The OPLS [optimized potentials for liquid simulations] potential functions for proteins, energy minimizations for crystals of cyclic peptides and crambin. *Journal of the American Chemical Society*, 110(6), pp.1657-1666.

Kaminski, G.A. et al., 2001. Evaluation and Reparametrization of the OPLS-AA Force Field for Proteins via Comparison with Accurate Quantum Chemical Calculations on Peptides †. *The Journal of Physical Chemistry B*, 105(28), pp.6474-6487.

Kauzmann, W., 1959. Some factors in the interpretation of protein denaturation. *Advances in Protein Chemistry*, 14, pp.1-63.

Kelley, N.W. et al., 2008. Simulating oligomerization at experimental concentrations and long timescales: A Markov state model approach. *The Journal of Chemical Physics*, 129(21), p.214707.

Khan, M., 2010. Interplay of protein misfolding pathway and unfolded-protein response in acute promyelocytic leukemia. *Expert Review of Proteomics*, 7(4), pp.591-600.

Kobilka, B.K., 2007. G protein coupled receptor structure and activation. *Biochimica Et Biophysica Acta*, 1768(4), pp.794-807.

Lampe, J.N. et al., 2010. Two-dimensional NMR and all-atom molecular dynamics of cytochrome P450 CYP119 reveal hidden conformational substates. *The Journal of Biological Chemistry*, 285(13), pp.9594-9603.

Lazaridis, T. & Karplus, Martin, 2003. Thermodynamics of protein folding: a microscopic view. *Biophysical Chemistry*, 100(1-3), pp.367-395.

Lin, M. et al., 2011. Constrained proper sampling of conformations of transition state ensemble of protein folding. *The Journal of Chemical Physics*, 134(7), p.075103.

Lindorff-Larsen, K. et al., 2010. Improved side-chain torsion potentials for the Amber ff99SB protein force field. *Proteins: Structure, Function, and Bioinformatics*, p.NA-NA.

Mackerell, A.D., Jr, 2004. Empirical force fields for biological macromolecules: overview and issues. *Journal of Computational Chemistry*, 25(13), pp.1584-1604.

MacKerell,, A.D. et al., 1998. All-Atom Empirical Potential for Molecular Modeling and Dynamics Studies of Proteins †. *The Journal of Physical Chemistry B*, 102(18), pp.3586-3616.

Mahoney, M.W. & Jorgensen, W.L., 2000. A five-site model for liquid water and the reproduction of the density anomaly by rigid, nonpolarizable potential functions. *The Journal of Chemical Physics*, 112(20), p.8910.

Marin, M., 2008. Folding at the rhythm of the rare codon beat. *Biotechnology Journal*, 3(8), pp.1047-1057.

McCammon, J.A., Gelin, B.R. & Karplus, M, 1977. Dynamics of folded proteins. *Nature*, 267(5612), pp.585-590.

Meirovitch, H., 2007. Recent developments in methodologies for calculating the entropy and free energy of biological systems by computer simulation. *Current Opinion in Structural Biology*, 17(2), pp.181-186.

Michels, P.C. & Clark, D S, 1997. Pressure-enhanced activity and stability of a hyperthermophilic protease from a deep-sea methanogen. *Applied and Environmental Microbiology*, 63(10), pp.3985-3991.

Nada, H. & van der Eerden, J.P.J.M., 2003. An intermolecular potential model for the simulation of ice and water near the melting point: A six-site model of H[sub 2]O. *The Journal of Chemical Physics*, 118(16), p.7401.

Nagaraj, N.S., Singh, O.V. & Merchant, N.B., 2010. Proteomics: a strategy to understand the novel targets in protein misfolding and cancer therapy. *Expert Review of Proteomics*, 7(4), pp.613-623.

Ozkan, S.B. et al., 2007. Protein folding by zipping and assembly. *Proceedings of the National Academy of Sciences of the United States of America*, 104(29), pp.11987-11992.

Patapati, K.K. & Glykos, N.M., 2011. Three force fields' views of the 3(10) helix. *Biophysical Journal*, 101(7), pp.1766-1771.

Pauling, L., Corey, R.B. & Branson, H.R., 1951. The structure of proteins: Two hydrogen-bonded helical configurations of the polypeptide chain. *Proceedings of the National Academy of Sciences*, 37(4), pp.205-211.

Petrov, D. & Zagrovic, B., 2011. Microscopic analysis of protein oxidative damage: effect of carbonylation on structure, dynamics, and aggregability of villin headpiece. *Journal of the American Chemical Society*, 133(18), pp.7016-7024.

Phillips, J.C. et al., 2005. Scalable molecular dynamics with NAMD. *Journal of Computational Chemistry*, 26(16), pp.1781-1802.

Plaxco, K.W., Simons, K.T. & Baker, D., 1998. Contact order, transition state placement and the refolding rates of single domain proteins. *Journal of Molecular Biology*, 277(4), pp.985-994.

Ponder, J.W. & Case, D.A., 2003. Force Fields for Protein Simulations. En *Advances in Protein Chemistry*. Elsevier, pp. 27-85. Available at: http://linkinghub.elsevier.com/retrieve/pii/S006532330366002X [Accedido febrero 7, 2012].

Price, D.J. & Brooks, Charles L, 3rd, 2002. Modern protein force fields behave comparably in molecular dynamics simulations. *Journal of Computational Chemistry*, 23(11), pp.1045-1057.

Privalov, P.L., 2009. Microcalorimetry of Proteins and Their Complexes. En J. W. Shriver, ed. *Protein Structure, Stability, and Interactions*. Totowa, NJ: Humana Press, pp. 1-39. Available at: http://www.springerlink.com/index/10.1007/978-1-59745-367-7_1 [Accedido febrero 7, 2012].

Rosas-Trigueros, J.L. et al., 2011. Insights into the structural stability of Bax from molecular dynamics simulations at high temperatures. *Protein Science: A Publication of the Protein Society*, 20(12), pp.2035-2046.

Schiene, C. & Fischer, G., 2000. Enzymes that catalyse the restructuring of proteins. *Current Opinion in Structural Biology*, 10(1), pp.40-45.

Schweiker, K.L. & Makhatadze, G.I., 2009. Protein Stabilization by the Rational Design of Surface Charge–Charge Interactions. En J. W. Shriver, ed. *Protein Structure, Stability, and Interactions*. Totowa, NJ: Humana Press, pp. 261-283. Available at: http://www.springerlink.com/index/10.1007/978-1-59745-367-7_11 [Accedido febrero 7, 2012].

Shea, J.E. & Brooks, C L, 3rd, 2001. From folding theories to folding proteins: a review and assessment of simulation studies of protein folding and unfolding. *Annual Review of Physical Chemistry*, 52, pp.499-535.

Shen, J.K., 2010. Uncovering specific electrostatic interactions in the denatured states of proteins. *Biophysical Journal*, 99(3), pp.924-932.

Shirts, M R & Pande, V S, 2001. Mathematical analysis of coupled parallel simulations. *Physical Review Letters*, 86(22), pp.4983-4987.

Shirts, Michael R. et al., 2003. Extremely precise free energy calculations of amino acid side chain analogs: Comparison of common molecular mechanics force fields for proteins. *The Journal of Chemical Physics*, 119(11), p.5740.

Silva, J.L. et al., 2010. Ligand binding and hydration in protein misfolding: insights from studies of prion and p53 tumor suppressor proteins. *Accounts of Chemical Research*, 43(2), pp.271-279.

Sonavane, S. & Chakrabarti, P., 2008. Cavities and Atomic Packing in Protein Structures and Interfaces C. Chothia, ed. *PLoS Computational Biology*, 4(9), p.e1000188.

Stumpe, M.C. & Grubmüller, H., 2009. Urea impedes the hydrophobic collapse of partially unfolded proteins. *Biophysical Journal*, 96(9), pp.3744-3752.

Suad, O. et al., 2009. Structural basis of restoring sequence-specific DNA binding and transactivation to mutant p53 by suppressor mutations. *Journal of Molecular Biology*, 385(1), pp.249-265.

Sundaramoorthy, M., Terner, J. & Poulos, T.L., 1995. The crystal structure of chloroperoxidase: a heme peroxidase--cytochrome P450 functional hybrid. *Structure (London, England: 1993)*, 3(12), pp.1367-1377.

Thomas, P.J., Qu, B.H. & Pedersen, P.L., 1995. Defective protein folding as a basis of human disease. *Trends in Biochemical Sciences*, 20(11), pp.456-459.

Tiberti, M. & Papaleo, E., 2011. Dynamic properties of extremophilic subtilisin-like serine-proteases. *Journal of Structural Biology*, 174(1), pp.69-83.

Uversky, V.N., 2009. Intrinsic disorder in proteins associated with neurodegenerative diseases. *Frontiers in Bioscience: A Journal and Virtual Library*, 14, pp.5188-5238.

van der Kamp, M.W. et al., 2010. Dynameomics: a comprehensive database of protein dynamics. *Structure (London, England: 1993)*, 18(4), pp.423-435.

Van Der Spoel, D. et al., 2005. GROMACS: Fast, flexible, and free. *Journal of Computational Chemistry*, 26(16), pp.1701-1718.

Vendruscolo, M. et al., 2001. Three key residues form a critical contact network in a protein folding transition state. *Nature*, 409(6820), pp.641-645.

Voter, A., 1998. Parallel replica method for dynamics of infrequent events. *Physical Review B*, 57(22), p.R13985-R13988.

Wang, R.Y.-R. et al., 2011. Modeling disordered regions in proteins using Rosetta. *PloS One*, 6(7), p.e22060.

Weiner, S.J. et al., 1984. A new force field for molecular mechanical simulation of nucleic acids and proteins. *Journal of the American Chemical Society*, 106(3), pp.765-784.

Wolynes, P.G., Onuchic, J.N. & Thirumalai, D., 1995. Navigating the folding routes. *Science (New York, N.Y.)*, 267(5204), pp.1619-1620.

Young, J.C. et al., 2004. Pathways of chaperone-mediated protein folding in the cytosol. *Nature Reviews. Molecular Cell Biology*, 5(10), pp.781-791.

Zhang, G., Hubalewska, M. & Ignatova, Z., 2009. Transient ribosomal attenuation coordinates protein synthesis and co-translational folding. *Nature Structural & Molecular Biology*, 16(3), pp.274-280.

In: Molecular Dynamics
Editors: D. E. Garcia and P. J. Green

ISBN: 978-1-62081-545-8
© 2012 Nova Science Publishers, Inc.

Chapter 2

INVESTIGATION OF STRUCTURAL PROPERTIES OF DRUG-METABOLIZING ENZYMES USING MOLECULAR DYNAMICS SIMULATION

Akifumi Oda[1,2,], Kana Kobayashi[1] and Ohgi Takahashi[1]*
[1]Faculty of Pharmaceutical Sciences,
Tohoku Pharmaceutical University, Japan
[2]Institute for Protein Research, Osaka University, Japan

ABSTRACT

Pharmacokinetic studies play an important role in drug design and development. Recently, many drug candidates have failed to be released into the market owing to pharmacokinetic problems. Thus, drug metabolism should be investigated in the early stages of drug design. Because a large number of drug candidates are screened in the early stages of drug design and development, fast and low-cost estimations of drug metabolism using computational approaches are useful. Structure-based drug design (SBDD), in which three-dimensional (3D) structures of proteins are used for molecular design, is a widely used computational approach. SBDD approaches can be used for drug target as well as for

---

* Correspondence concerning this article should be addressed to Akifumi Oda. E-mail: oda@tohoku-pharm.ac.jp; Fax: +81-22-275-2013.

drug-metabolizing enzymes in drug design testing. For example, undesirable inhibition of drug metabolism can be identified and the effects of genetic polymorphism can be predicted for drug candidates using the SBDD approach for drug-metabolizing enzymes. Although 3D structures of proteins are indispensable to the SBDD approach, experimental structures of many drug-metabolizing enzymes are still unknown. Thus, computational procedures for prediction and refinement of the 3D structures of drug-metabolizing enzymes are important. In this chapter, structural refinements of UDP-glucuronyltransferase using molecular dynamics simulations are described. Because there are polymorphisms for this enzyme, its detailed structure is essential for SBDD. Thus, molecular dynamics simulations are expected to be useful for structural refinements of drug-metabolizing enzymes.

1. INTRODUCTION

1.1. Computational Studies for Drug Metabolizing Enzymes

In recent times, pharmacokinetic studies have had important roles in the early stages of drug design and development [1]. Many drug candidates have failed to be released into the market owing to pharmacokinetic problems, such as drug interactions and genetic polymorphism of drug-metabolizing enzymes [2]. Therefore, drug metabolism should be examined in the early stages of drug design. Because a large number of compounds are tested in the early stages of drug design and development, fast and low-cost estimations of drug metabolism are indispensable to efficient drug design trials. In this respect, computational approaches for investigating drug metabolisms, e.g., virtual screening and quantitative structure–activity relationship (QSAR) studies appear to be useful in reducing the number of experiments and the associated monetary and environmental costs [3]. Structure-based drug design (SBDD), in which three-dimensional (3D) structures of proteins are used for molecular design, is one of the most widely used computational approaches for drug design and development. SBDD approaches can be used not only for drug targets but also for drug-metabolizing enzymes to investigate the pharma cokinetic aspects of drug candidates. For example, undesirable inhibition of drug metabolism can be identified and the effects of genetic polymorphisms can be predicted for drug candidates using the SBDD approach. Although 3D structures of proteins are indispensable to the SBDD approach, experimental structures are still unknown for many drug-metabo-lizing enzymes. Thus, computational procedures for the prediction and refinement of 3D structures of these enzymes are important.

Although the use of computational methods for investigating pharmacokinetics remains insufficient, some structural bioinformatics researches have been performed for drug-metabolizing enzymes. Cytochrome P450 (CYP) is one of the most widely investigated enzymes by computational methods. For CYP enzymes, computational studies that use structural bioinformatics and computational chemical methods [4] as well as QSAR [5] have been reported, and reliable results can be obtained. In contrast, computational investigations for other drug-metabolizing enzymes have been insufficient, and the computational predictions of drug metabolisms by these enzymes are still challenging projects. *N*-acetyltransferase 2 (NAT2), which catalyzes *N*-acetyl conjugation of allylamine derivatives, is a drug-metabolizing enzyme. We predicted the 3D structure of NAT2 with acetyl-cofactor A using molecular dynamics (MD) simulations [6]. Therefore, structural bioinformatics investigation of NAT2 can be performed. Another drug-metabolizing enzyme, UDP-glucuronyltransferase (UGT), catalyzes the glucuronide conjugation of several drug compounds. Although UGT is one of the most important phase II drug-metabolizing enzymes, the 3D-structure of human UGT is not completely known. Only the 3D structure of human UGT2B7 is known; however, the N-terminal region of UGT2B7 has not been experimentally determined. Because reliable computational model structures have also not been predicted for UGTs, SBDD strategies for computational investigations of drug metabolisms by UGT cannot be used. Genetic polymorphisms of UGTs have been reported [7-8]. Pharmacokinetic studies of drug candidates that include investigations of polymorphic enzymes contribute significantly to personalized medicinal strategies. As mentioned above, computational studies of biomolecules related to pharmacokinetics are a cutting-edge area of drug design and development trials, particularly for drug-metabolizing enzymes other than CYP. The 3D structures of several enzymes have not been determined by experimental studies and have not been predicted by computational studies. In this chapter, the 3D structures of a polymorphic drug-metabolizing enzyme, UGT1A1, were predicted and structural refinements were performed by MD simulations.

1.2. UGT1A1

UGT metabolizes several endogenous and exogenous compounds [9]. UGT is localized in the endoplasmic reticulum membrane of living cells, such as stem cells, and it includes UDP-glucuronic acid (UDPGA) as a cofactor.

UGT catalyzes glucuronide conjugation, in which glucuronic acid moieties are added to several compounds. In general, the compounds that are conjugated with glucuronic acids have characteristic functional groups, such as hydroxyl, carboxyl, amino, and/or thiol groups, and the glucuronic acid moieties attach to these functional groups. For example, glucuronide conjugation of a compound with a hydroxyl group is as follows:

$$ROH + UDPGA \rightarrow R\text{-}O\text{-glucuronate} + UDP$$

Because of the high solubility of the glucuronate moiety, the solubility of the compound conjugated with glucuronic acid is increased, and the conjugated compounds are easily excreted.

Although UGT has some isozymes, two types of isozymes, UGT1 and UGT2 families, have been reported for human UGT [10]. Both UGT1 and UGT2 family isozymes include approximately 500 amino acid residues, and the amino acid sequences of the C-terminal regions are homologous among the isozymes. In contrast, the amino acid sequences of the N-terminal regions of UGTs are different. The C-terminal and N-terminal regions are considered to be the UDPGA-binding domain and ligand-recognition domain, respectively. All UGTs include UDPGA as a cofactor, and the amino acid sequence of the UDPGA-binding domain is conserved. Because different isozymes recognize different substrates, various amino acid sequences of the ligand-recognition domains of UGTs have been observed. In general, a greater variety of substrates are metabolized by UGT1 family isozymes than that by UGT2 isozymes. UGT metabolizes several drugs and drug-related compounds. SN-38, which is an active metabolite of an anticancer drug, CTP-11 (irinotecan), is metabolized by UGT1 isozymes, such as UGT1A1 and UGT1A7. The glucuronide conjugation of acetaminophen is mainly catalyzed by UGT1A6. Morphine and 3'-azido-3'-deoxythymidine are conjugated with glucuronates by UGT2B7 [11].

For several UGT isozymes, many genetic mutants have been found, and the genetic polymorphisms of UGTs cause genetic diseases and individual differences in drug efficacies and/or drug metabolisms. For example, Crigler–Najjar syndrome type I and type II and Gilbert's syndrome, all of which cause bilirubin metabolism disorder, result from deficiency of UGT1A1 or substitution of amino acid residues in this isozyme. Thus, UGT1A1 plays an important role in bilirubin metabolism [12]. In addition, genetic polymorphism of UGT1A1 was found in patients with infantile jaundice in Japanese population [13], and it has been reported to influence the incidence of breast cancer as well [14]. Furthermore, UGT1A1 catalyzes the glucuronide conjugation of several drugs and drug-related compounds, such as SN-38, which is an active metabolite of an anticancer drug, CTP-11; thus, genetic polymorphism of UGT1A1 plays an important role in drug metabolism.

As mentioned above, UGTs are significant enzymes in metabolism of endogenous and exogenous compounds; therefore, investigation of the structural features of these enzymes and the influence of polymorphism on their 3D structures contribute significantly to drug metabolism studies. However, only a few 3D structures of UGT enzymes have been determined experimentally because UGT is one of the membrane-binding proteins, which are difficult to crystallize. For human UGT, only the experimental structure of the ligand-free state of UGT2B7 has been reported. However, only the 3D structure of the UDPGA-binding domain at the C-terminal region was determined, and the ligand-recognition domain at the N-terminal region is still unknown even for UGT2B7 [15]. In addition, experimental 3D structures of UGT1A1, which plays an important role in drug metabolism, and all other UGT1 family isozymes are unknown. Computational approaches for predicting 3D structures of UGT enzymes have been undertaken; e.g., computational modeling using protein structure prediction [16] and 3D-QSAR studies [17-19]. Although some structural features of UGT have been elucidated by these studies, detailed structures still have not been obtained for UGT enzymes. 3D-QSAR cannot predict whole protein structures because it has a ligand-based approach. The 3D structures obtained by structural bioinformatics approach [16] were not accurate because a vegetal UGT structure was used as the template for protein modeling of human UGT and UGT2B7 structure was not used as the template.

For the structural prediction of human UGT1A1, the human UGT2B7 structure was used as one of the templates for UGT1A1, and more accurate protein modeling than that used in other modeling trials was performed. For drug-metabolizing enzymes, including UGT and others, many isozymes are frequently included, and genetic polymorphisms are frequently reported. Thus, many proteins have to be considered when investigation of one drug-metabolizing enzyme is performed. Experimental determination of 3D structures of all proteins is impractical because of extremely high costs, and computational approaches, e.g., structural predictions and refinements, are important in research on polymorphic drug-metabolizing enzymes. Recently, several methods to predict protein 3D structures have been developed, and useful predicted structures were obtained from a metaserver using consensus methods with multiple procedures, and the final predicted structure was determined by a majority vote [20]. In this study, a metaserver was used for the prediction of the 3D structure of UGT1A1. For one of the UGT1 family enzymes, UGT1A6, a disulfide bond between Cys121 and Cys125 was important for enzymatic glucuronide conjugation [21]. Whereas for UGT1A1, a disulfide bond of Cys127, which is the corresponding residue to Cys125 in UGT1A6, was reported to be unimportant in enzyme activity [22]. For the predicted structure of UGT1A1, we investigated the structural environment of Cys127 and elucidated the role of the disulfide bond in UGT1A1 using molecular modeling and MD simulations.

2. Methods

In this chapter, prediction of the 3D structure of UGT1A1 from the amino acid sequence is described. For structural prediction, the protein homology/ analogy recognition engine (Phyre) [23-25], which is the metaserver that uses the fold recognition method, was used. In Phyre, all steps of the prediction, from secondary structure prediction to construction of the 3D structure, are performed fully automatically; therefore, template protein structures are also automatically identified. For UGT1A1, protein structures whose protein data bank (PDB) identifiers are 2IYA, 2IYF, 2PQ6, 2C1Z, 2ACV, 2VCU, 1IIR, 1RRV, 2O6L, and 2P6P were used as templates. These structures were derived from RCSB PDB [26]. Structure 2O6L is the 3D structure of the C-terminal domain of UGT2B7, and the other structures are the structures of transferases. From the Phyre-predicted structure of UGT1A1, two types of structural models with and without a disulfide bond related to Cys127 were constructed. One model had a disulfide bond between Cys127 and one other Cys, and another model had the thiol form of Cys127. The template structures used in this study were X-ray crystallographic structures, thus, these structures have no hydrogen atoms. Therefore, all hydrogens were added for the constructed UGT1A1 models. These modeling procedures were performed using tLeap module of AMBER 9 [27].

For the constructed models, structural refinements were performed using molecular mechanical (MM) calculations and MD simulations. The calculations were performed using explicit water boxes. The box sizes were determined by the protein surfaces with a margin of 8 Å, and TIP3P [28] was used as the water model. The sodium ions were added as counter ions to electrically neutralize whole systems. For the MM and MD calculations, the AMBER ff99SB force field [29-30], which is one of the most widely used classical molecular force fields, was used in this study. In the structural optimization steps that used MM calculations, only water molecules were optimized at first, and then the whole system was optimized. The maximum computational steps for water and whole-system optimizations were set to 5,000 and 10,000 cycles, respectively. The convergence conditions in terms of energy gradients were less than 1.0×10^{-1} kcal/mol Å and less than 1.0×10^{-4} kcal/mol Å for water and whole-system optimizations, respectively. These optimizations were performed under constant volume conditions. For the optimized UGT1A1 structures with and without a disulfide bond, 30-ps temperature-increase simulations with a 0.5-fs time step were performed from 0 K to 300 K. Only the water molecules were moved for the temperature-increase MD simulations. Next, 2-ns MD simulations were performed with a time step of 1 fs at 300 K. The temperature-increase MD and equilibrating MD simulations were performed under constant volume and pressure conditions, respectively. For both MD simulations, the SHAKE method [31] was adopted, and periodic boundary conditions were used. The particle mesh Ewald method was used for the calculations of electrostatic interactions, and the cut-off for nonbonding terms was 10 Å. The MM and MD calculations were performed using AMBER 9 [27].

For the refined UGT1A1 structures, ligand-binding sites were searched, and structural features of the models were investigated. The ligand-binding site searches were performed using the Q-SiteFinder web server [32-33]. Because Q-SiteFinder has a routine of adding hydrogen atoms for proteins, hydrogens of the constructed UGT1A1 models were deleted. The ligand-binding site searches were performed for three types of UGT1A1 model: 1) the unchanged model constructed by Phyre before MM and MD calculations, 2) the model with the disulfide bond after MM and MD structural refinements, and 3) the model without the disulfide bond after MM and MD refinements.

3. RESULTS AND DISCUSSION

The UGT1A1 structure predicted by the Phyre metaserver is shown in Figure 1. Structural predictions were performed both for the C-terminal UDPGA-binding region for which amino acid sequences are comparatively conserved and for the N-terminal region. Although 3D structures of 28 residues at the N-terminal could not be determined using the fold recognition method, the structures of important residues for enzymatic activities [22] were obtained by Phyre. In addition, the N-terminal and C-terminal domains were in contact with each other, and glucuronic conjugations of substrates by UDPGA appeared to be catalyzed by UGT1A1 in the boundary region.

Table 1. Predicted ligand binding sites

	Rank of site	Residues included in sites	Vol/Å^3
Phyre structure	Site 1 (N-terminal)	T150, D151, P152, F153, L154, P155, P158, F170, F171, L172, L175, P176, C177, S178, E182, Y230, D246, L247, L248, S249, S250, A251, W254, F256, I268, M269, P270, N271, M272	595
	Site 2 (N-terminal)	I33, P34, V35, D36, P62, Y79, P80, V81, P82, F83, Q84, M122, L123, L124, S125, G126, C127, S128, H129, L130, L131, L136, F153, L154, P155, C156	357
	Site 3 (boundary)	S38, L41, N279, H282, S306, L307, G308, R336, Y337, W354, L355, P356, Q357, N358, D359, F369, T371, H372, G374, H376, G377, E380	326
without a disulfide bond	Site 1 (N-terminal)	L123, L124, P152, F153, P155, F170, F171, L172, H173, A174, L175, P176, C177, S178, E182, F190, L247, L248, S249, S250, A251, W254, F256, I268, M269, M272	513
	Site 2 (boundary)	F83, Q84, T112, K114, K115, I116, D119, S120, P187, N188, P189, M310, L393, F394, G395, D396	334
	Site 3 (N-terminal)	P158, I159, A161, Q162, S165, L166, P167, T168, V169, F170, R240, D246, L247, S250, A251, S252, V253	261
with a disulfide bond	Site 1 (boundary)	P34, V35, D36, G37, S38, W40, L41, P62, A64, S65, L66, Y67, I68, E86, H282, G308, S309, R336, K353, W354, L355, Q357, H376	422
	Site 2 (boundary)	L175, P176, C177, E182, W254, S258, D259, F260, D263, P265, R266, I268, F274, Y379, D396, D399, N400, R403, M404, K407	386
	Site 3 (boundary)	D36, H39, F83, Q84, R85, E86, D87, Q107, R108, I110, K111, T112, S309, M310, V311, I314, R336, Y337, K353	452

The predicted 3D position of Cys127, which may form the disulfide bond [22], is illustrated in Figure 1. As shown in the figure, Cys127 is located adjacent to Cys156, and the distance between the sulfur atoms of Cys127 and Cys156 is 5.84 Å. Although the distance was too large for the formation of a disulfide bond, Cys156 is the most possible residue that would form a disulfide bond with Cys127, if Cys127 does indeed form a disulfide bond. Thus, two types of models, with and without a disulfide bond between Cys127 and Cys156, were constructed, and optimizations using MM calculations and structural refinements using MD simulations were performed for the two models.

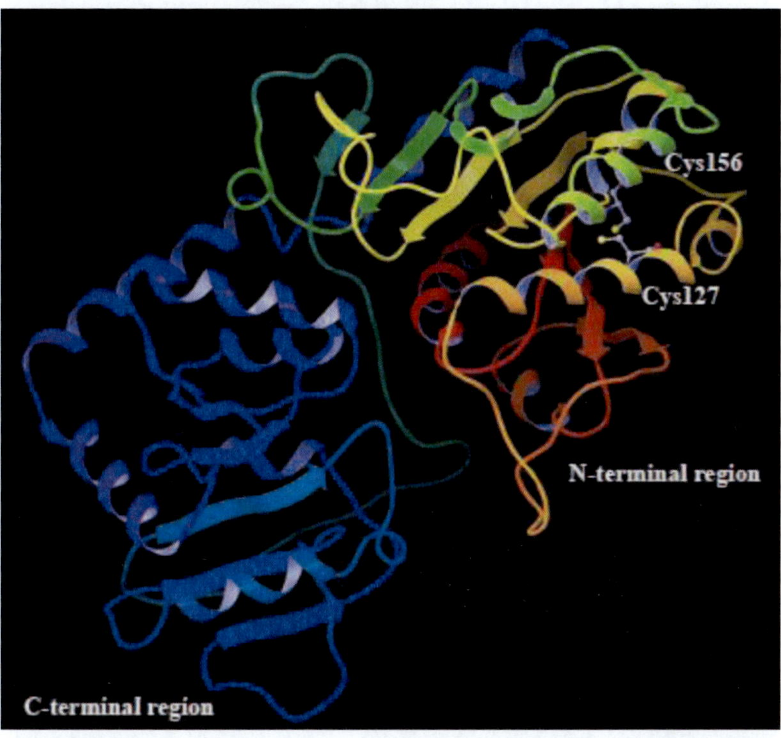

Figure 1. The predicted structure of UGT1A1 using Phyre. Cys127 and Cys156 are illustrated by ball-and-stick models.

In Figure 2, the root mean square deviations (RMSDs) for MD trajectories are illustrated. The blue line indicates the model without a disulfide bond, and the magenta line indicates the model with a disulfide bond. The initial structures of MD simulations were used as the reference structures for RMSD calculations. As shown in the figure, MD simulations appear to converge at 2 ns. Although the structures were not greatly changed in both the models using MD simulation, the RMSD values for the model with a disulfide bond were slightly larger than those for the model without a disulfide bond. The 3D structures of UGT1A1 with and without a disulfide bond after structural refinements are shown in Figure 3. The structure of the model with a disulfide bond is shown as a brown ribbon and that of the model without a disulfide bond is shown as a light-blue ribbon. RMSD between the two models was 3.706 Å.

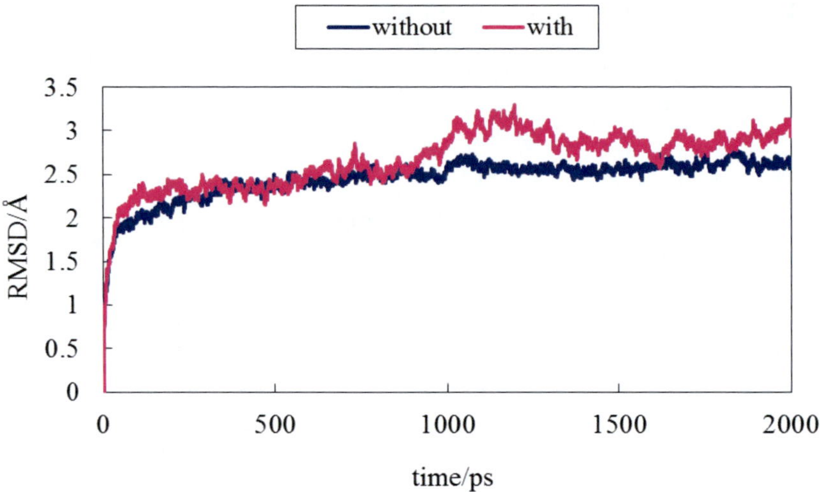

Figure 2. RMSDs with MD simulations. The blue line indicates RMSDs for the model without a disulfide bond, and the magenta line indicates RMSDs for the model with a disulfide bond.

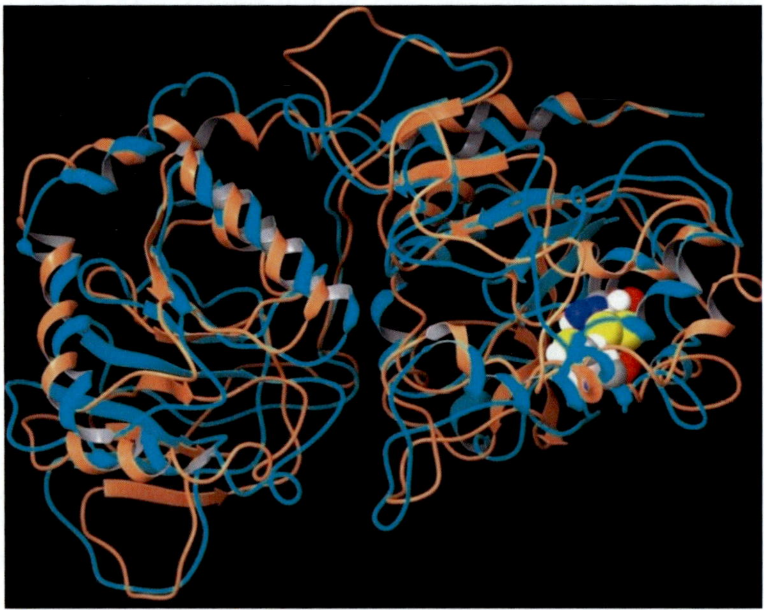

Figure 3. Refined UGT1A1 structures. The models with and without a disulfide bond are shown as brown and light blue ribbons, respectively. Cys127 and Cys156 are illustrated by space-filling models.

The candidates for the ligand-binding site of UGT1A1 predicted by Q-SiteFinder are shown in Table 1. In the table, the site residues and volumes of the pockets are indicated. The pocket candidates were ranked by Q-SiteFinder in order of the potency of ligand binding, and the top three candidates are shown in the table. Site 1 is the most probable ligand-binding site candidate, and Site 2 and Site 3 are the second- and third-best candidates, respectively. In the parentheses following the site names, the location of the sites are described; i.e., the N-terminal domain, C-terminal domain, and the boundary region. In the Phyre structure row, the Q-SiteFinder results for unchanged structures before MM and MD calculations are shown. In the without a disulfide bond and with a disulfide bond rows, the results for the models after MM and MD structural refinements without and with a disulfide bond are illustrated, respectively. As shown in the table, the candidates were found in the N-terminal domain and boundary region of the Phyre structure and the model without a disulfide bond, although the candidates were identified only in the boundary region for the model with a disulfide bond. The site volume of Site 1 for the model with a disulfide bond was smaller than that for the Phyre structure and the model without a disulfide bond. These results indicate that the 3D structure of the model with a disulfide bond is different from the Phyre structure and the model without a disulfide bond. These results are consistent with RMSDs shown in Figure 2. Because of the small volume of Site 1 of the model with a disulfide bond and the long distance between Cys127 and Cys156 in the Phyre structure, the model without a disulfide bond may be more reasonable for ligand binding. This is consistent with the experimental results reported previously [22] in which the disulfide bond of Cys127 was reported to be unimportant for enzymatic activity of UGT1A1. In the previous report [22], the disulfide bond for all Cys was described as unimportant. In our UGT1A1 model, the distances between pairs of Cys were larger than 10 Å except for the distance between Cys127 and Cys156. In addition, the distance between Cys127 and Cys156 was 5.86 Å, as mentioned above. Because all Cys–Cys distances were too long to form a disulfide bond, our computational models appear to reasonably explain the experimental results, and our model is expected to be useful for pharmacokinetic studies of UGT1A1.

ACKNOWLEDGMENTS

This work was supported by a grant-in-aid for scientific research (23790137) from the Japan Society for the Promotion of Science.

REFERENCES

[1] van de Waterbeemd, H.; Gifford, E. (2003). ADMET In silico modelling: towards prediction paradise. *Nat. Rev. Drug Discovery, 2*, 192-204.

[2] Kola, I.; Landis, J. (2004). Can the pharmaceutical industry reduce attrition rates? *Nat. Rev. Drug Discovery, 3*, 711-715.

[3] Leach, A. R. (2001). *Molecular Modelling, 2nd ed.*, Essex, UK, Pearson Education Limited.

[4] Oda, A.; Yamaotsu, N.; Hirono, S. (2004). Studies of binding modes of (S)-mephenytoin to wild types and mutants of cytochrome P450 2C19 and 2C9 using homology modeling and computational docking. *Pharm. Res., 21*, 2270-2278.

[5] Locuson, C. W. II; Suzuki, H.; Rettie, A. E.; Jones, J. P. (2004). Charge and substituent effects on affinity and metabolism of benzbromarone-based CYP2C19 inhibitors. *J. Med. Chem., 47*, 6768-6776.

[6] Oda, A.; Kobayashi, K.; Takahashi, O. (2010). Computational study of the 3D structure of *N*-acetyltransferase 2–acetyl coenzyme A complex. *Biol. Pharm. Bull., 33*, 1639-1643.

[7] Nagar, S.; Remmel, R. P. (2006). Uridine diphosphoglucuronosyltransferase pharmacogenetics and cancer. *Oncogene, 25*, 1659-1672.

[8] Beutler, E.; Gelbart, T.; Demina, A. (1998). Racial variability in the UDP-glucuronosyltransferase 1 (UGT1A1) promoter: a balanced polymorphism for regulation of bilirubin metabolism? *Proc. Natl. Acad. Sci. USA, 95*, 8170-8174.

[9] Radominska-Pandya, A.; Czernik, P. J.; Little, J. M.; Battaglia, E.; Mackenzie, P. I. (1999). Structural and functional studies of UDP-glucuronosyltransferases. *Drug Metab. Rev., 31*, 817-899.

[10] Mackenzie, P. I.; Owens, I. S.; Burchell, B.; Bock, K. W.; Bairoch, A.; Bélanger, A.; Fournel-Gigleux, S.; Green, M.; Hum, D. W.; Iyanagi, T.; Lancet, D.; Louisot, P.; Magdalou, J.; Chowdhury, J. R.; Ritter, J. K.; Schachter, H.; Tephly, T. R.; Tipton, K. F.; Nebert, D. W. (1997). The UDP glycosyltransferase gene superfamily: recommended nomenclature update based on evolutionary divergence. *Pharmacogenetics, 7*, 255-269.

[11] Barbier, O.; Turgeon, D.; Girard, C.; Green, M. D.; Tephly, T. R.; Hum, D. W.; Bélanger, A. (2000). 3'-azido-3'-deoxythimidine (AZT) is glucuronidated by human UDP-glucuronosyltransferase 2B7 (UGT2B7). *Drug. Metab. Dispos., 28*, 497-502.

[12]   Monaghan, G.; Ryan, M.; Seddon, R.; Hume, R.; Burchell, B. (1996). Genetic variation in bilirubin UDP-glucuronosyltransferase gene promoter and Gilbert's syndrome. *Lancet, 347,* 578-581.

[13]   Akaba, K.; Kimura, T.; Sasaki, A.; Tanabe, S.; Wakabayashi, T.; Hiroi, M.; Yasumura, S.; Maki, K.; Aikawa, S.; Hayasaka, K. (1999). Neonatal hyperbilirubinemia and a common mutation of the bilirubin uridine diphosphate-glucuronosyltransferase gene in Japanese. *J. Hum. Genet., 44,* 22-25.

[14]   Guillemette, C.; Millikan, R. C.; Newman, B.; Housman, D. E. (2000). Genetic polymorphisms in uridine diphospho-glucuronosyltransferase 1A1 and association with breast cancer among African Americans, *Cancer Res., 60,* 950-956.

[15]   Miley, M. J.; Zielinska, A. K.; Keenan, J. E.; Bratton, S. M.; Radominska-Pandya, A.; Redinbo, M. R. (2007). Crystal structure of the cofactor-binding domain of the human phase II drug-metabolism enzyme UDP-glucuronosyltransferase 2B7. *J. Mol. Biol., 369,* 498-511.

[16]   Locuson, C. W.; Tracy, T. S. (2007). Comparative modelling of the human UDP-glucuronosyltransferases: Insights into structure and mechanism. *Xenobiotica, 37,* 155-168.

[17]   Sorich, M. J.; Smith, P. A.; McKinnon, R. A.; Miners, J. O. (2002). Pharmacophore and quantitative structure activity relationship modelling of UDP-glucuronosyltransferase 1A1 (UGT1A1) substrates, *Pharmacogenetics, 12,* 635-645.

[18]   Ethell, B. T.; Ekins, S.; Wang, J.; Burchell, B. (2002). Quantitative structure activity relationships for the glucuronidation of simple phenols by expressed human UGT1A6 and UGT1A9. *Drug Metab. Dispos., 30,* 734-738.

[19]   Smith, P. A.; Sorich, M. J.; Low, L. S. C.; McKinnon, R. A.; Miners, J. O. (2004). Towards integrated ADME prediction: past, present and future directions for modelling metabolism by UDP-glucuronosyl-transferases. *J. Mol. Graph. Model., 22,* 507-517.

[20]   Park, S.; Chikenji, G.; Hirokawa, T.; Tomii, K.; Takada, S. (2005). Present and future prospects of protein structure prediction. *Transactions of Japanese Society Artificial Intelligence, 20,* 479-485.

[21]   Ikushiro, S.; Emi, Y.; Iyanagi T. (2002). Activation of glucuronidation through reduction of a disulfide bond in rat UDP-glucuronosyl-transferase 1A6. *Biochemistry, 41,* 12813-12820.

[22] Ghosh, S. S.; Lu, Y.; Lee, S. W.; Wang, X.; Guha, C.; Roy-Chowdhury, J.; Roy-Chowdhury, N. (2005). Role of cysteine residues in the function of human UDP glucuronosyltransferase isoform 1A1 (UGT1A1). *Biochem. J.*, *392*, 685-692.

[23] Bennett-Lovsey, R. M.; Herbert, A. D.; Sternberg, M. J. E.; Kelley, L.A. (2008). Exploring the extremes of sequence/structure space with ensemble fold recognition in the program Phyre. *Proteins, Struct. Funct. Bioinf.*, *70*, 611-625.

[24] *Imperial College Phyre Server*, <http://www.sbg.bio.ic.ac.uk/~phyre/>.

[25] Kelley, L. A.; Sternberg, M. J. E. (2009). Protein structure prediction on the web: a case study using the Phyre server. *Nat. Protocols*, *4*, 363-371.

[26] Berman, H. M.; Westbrook, J.; Feng, Z.; Gilliland, G.; Bhat, T. N.; Weissig, H.; Shindyalov, I. N.; Bourne, P. E. (2000). The protein data bank. *Nucleic Acids Res.*, *28*, 235-242.

[27] Case, D. A.; Darden, T. A.; Cheatham, T. E., III; Simmerling, C. L.; Wang, J.; Duke, R. E.; Luo, R.; Merz, K. M.; Pearlman, D. A.; Crowley, M.; Walker, R. C.; Zhang, W.; Wang, B.; Hayik, S.; Roitberg, A.; Seabra, G.; Wong, K. F.; Paesani, F.; Wu, X.; Brozell, S.; Tsui, V.; Gohlke, H.; Yang, L.; Tan, C.; Mongan, J.; Hornak, V.; Cui, G.; Beroza, P.; Mathews, D. H.; Schafmeister, C.; Ross, W. S.; Kollman, P. A. (2006). *AMBER9*, San Francisco, CA: University of California.

[28] Jorgensen, W. L.; Chandrasekhar, J.; Madura, J.; Klein, M. L. (1983). Comparison of simple potential functions for simulating liquid water. *J. Chem. Phys.*, *79*, 926-935.

[29] Wang, J.; Cieplak, P.; Kollman, P. A. (2000). How well does a restrained electrostatic potential (RESP) model perform in calculating conformational energies of organic and biological molecules? *J. Comput. Chem.*, *21*, 1049-1074.

[30] Hornak, V.; Abel, R.; Okur, A.; Strockbine, B.; Roitberg, A.; Simmerling, C. (2006). Comparison of multiple Amber force fields and development of improved. *Proteins, Struct. Funct. Bioinf.*, *65*, 712-725.

[31] Ryckaert, J.-P.; Ciccotti, G.; Berendsen, H. J. C. (1977). Numerical integration of the cartesian equations of motion of a system with constraints: Molecular dynamics of *n*-alkanes. *J. Comput. Phys.*, *23*, 327-341.

[32] Laurie, A. T. R.; Jackson, R. M. (2005). Q-SiteFinder: an energy-based method for the prediction of protein–ligand binding sites. *Bioinformatics, 21,* 1908-1916.

[33] Laurie A. T. R.; Jackson R. M. *Q-SiteFinder: Ligand Binding Site Prediction Server,* <http://bmbpcu36.leeds.ac.uk/qsitefinder/>.

In: Molecular Dynamics
Editors: D. E. Garcia and P. J. Green

ISBN: 978-1-62081-545-8
© 2012 Nova Science Publishers, Inc.

Chapter 3

DOUBLE-PULSE LASER CONTROL OF ULTRAFAST OPTICAL KERR EFFECT IN LIQUID

V. G. Nikiforov[*] and V. S. Lobkov

Kazan Physical-Technical Institute, Kazan Scientific Center, Russian
Academy of Sciences, Sibirskii trakt, Kazan, Russia

ABSTRACT

The optical control of the molecular motions in the neat liquids of carbon tetrachloride CCl_4 and chloroform $CHCl_3$ at room temperature through the non-resonant excitation was enhanced by means of the double-pulse pump-probe technique. When the separation time of the pump pulses and their relative intensity were varied, the amplification or the cancellation of the coherent vibrations of the molecules was achieved. The molecular responses were detected by the time-resolved optically heterodyne-detected optical-Kerr-effect (OKE) technique. It was shown that the time-resolved polarization selective spectroscopy of the molecular vibrational and rotational dynamics in liquid can be based on the double-pulse laser control.

---

[*] E-mail: vgnik@mail.ru.

INTRODUCTION

The laser control of the molecular dynamics is a subject of the intensive theoretical and experimental studies. Nowadays, the femtosecond techniques [1] are commonly used to implement the laser control. The fundamental problems, such as the controlling of the chemical reactions [2-4], the energy transfer in the photosynthetic complexes [5,6], the selective photoionization of the molecular isotopes [7], the selective photodissociation and the creation of new bonds in the molecules [8], the photoisomerization of the molecules [9] etc are solving by the femtosecond technique.

The multiple-pulse technique is one of the laser control methods. The wave packets in the medium are formed by the timed sequences of the femtosecond pulses, so the time delay between pulses is one of the key parameter. For example the mode-selective amplification in α-perylene molecular crystals was enhanced by this technique [10]. The double-pulse and four-pulse excitations were implemented to control the coherent phonon oscillations in various solid materials [11-18].

We use the double-pulse non-resonant excitation to control the amplitudes of the individual modes in the vibrational responses of the molecules in liquid. We suppose that by the short relaxation time of the coherent molecular motions (of less or a few picoseconds) the technique was not employing in liquid at room temperature. In the experiments the controlling parameters are the relative intensities of the pump pulses and the time delay between them. The molecular dynamics is probed by the time-resolved optically heterodyne-detected optical Kerr effect (OKE) [19-21]. For the first time the one-pulse excitation was proposed to probe the orientational dynamics of the molecules in liquid [22, 23]. The femtosecond laser pulses excite both picosecond orientational responses and femtosecond responses of the coherent molecular vibrational motions.

THEORETICAL TREATMENT

The interpretation of the experimental data with the single-pulse excitation requires the simulation of the third-order optical response of the medium. The double-pulse excitation requires the simulation of the fifth-order optical response. There are several theoretical treatments to analyze the OKE signal [24-27]. We use the simulation based on the model in [28, 29] to decompose the total OKE signal into the component molecular responses at the single- and double-pulse excitation of the medium. The description of the treatment is as follows. The OKE signal $S(\tau)$ detected in the experiment is the time-averaged optical response of the medium:

$$S(\tau) \propto \int_{-\infty}^{\infty} s(\tau,t)\,dt ,$$

(1)

where $\blacklozenge$ is the delay between the first pump pulse and the probe pulse. The function $s(\tau,t)$ includes the molecular responses $R(t)$:

$$s(\tau,t) = E_{lo}(t-\tau)E_0(t-\tau)\{R_{el}(t) + R_{osc}(t) + R_{or}(t) + R_{lib}(t)\}.$$

(2)

The pump pulse $E_p(t)$, the probe pulse $E_0(t)$ and the pulse of the local oscillator field $E_{lo}(t)$ are of the duration τ_{pul} (FWHM). The envelope of the intensities is simulated by the $I_p(t) \propto I_0(t) \propto I_{lo}(t) \propto \operatorname{sech}(t/(0.38 \cdot \tau_{pul}))$ function. If we use the double-pulse excitation the expression for $I_p(t)$ consists of the two terms:

$$I_p(t) = I_p^{(1)} \operatorname{sech}(t/(0.38 \cdot \tau_{pul})) + I_p^{(2)} \operatorname{sech}((t-\tau_{12})/(0.38 \cdot \tau_{pul})),$$

(3)

where $I_p^{(i)}$ is the intensity of the i-th pump pulse, τ_{12} is the delay between the pump pulses.

The instantaneous electronic response is proportional to the coefficient of the electronic cubic hyperpolarizability γ and the intensity of the pump pulses:

$$R_{el}(t) \propto \gamma I_p(t).$$

(4)

The response function of the intramolecular vibrations $R_{vib}(t)$ is:

$$R_{vib}(t) \propto \sum_{i=1}^{N} \left(\alpha_{vib}^{i}\right)^{2} \Phi_{vib}(\Omega_{vib}^{i}, \tau_{vib}^{i}, t) \tag{5}$$

$$\Phi_{vib}(\Omega_{vib}, \tau_{vib}, t) = \left(\Omega_{vib}^{2} - \tau_{vib}^{-2}\right)^{-1/2} \int_{0}^{\infty} I_{p}(t-t') \cdot \exp\left(-\frac{t'}{\tau_{vib}}\right) \cdot \sin\left[\left(\Omega_{vib}^{2} - \tau_{vib}^{-2}\right)^{1/2} \cdot t'\right] \cdot dt'$$

where N is the number of the vibrational modes excited by the pump pulses; α_{vib}^{i} is the coefficient characterizing the change of the molecular polarizability at the excitation of the i-th mode; Ω_{vib}^{i} and τ_{vib}^{i} are the frequency and the relaxation time of the i-th mode respectively.

If the molecules have the anisotropy of the polarizability $\Delta\alpha \neq 0$ the laser pulse induces the orientational anisotropy of the molecules relaxed by the rotational diffusion mechanism. The orientational response describes by the following expression:

$$R_{or}(t) \propto \Delta\alpha^{2} \int_{0}^{\infty} I_{p}(t-t') \exp\left(-\frac{t'}{\tau_{or}}\right) dt' \tag{6}$$

where τ_{or} is the relaxation time of the orientational response.

In addition to the orientational anisotropy of molecules, librations of the molecules with the frequency Ω_{lib} are excited as well. The description of the librational response is based on the distribution function of the librational frequencies $\rho(\Omega_{lib})$ reflected the local inhomogeneity of the medium:

$$R_{lib}(t) \propto \Delta\alpha^{2} \int_{0}^{\infty} d\Omega_{lib} \rho(\Omega_{lib}) \Phi_{lib}(\Omega_{lib}, t) \tag{7}$$

$$\Phi_{lib}(\Omega_{lib}, t) = \left(\Omega_{lib}\right)^{-1} \int_{0}^{\infty} I_{p}(t-t') \cdot \sin\left[\Omega_{lib} \cdot t'\right] \cdot dt'$$

The results in [29] give reason for Maxwell distribution as the distribution function of the librational response in particular for the chloroform:

$$\rho(\Omega_{lib}) \propto (\Omega_{lib})^2 \exp\left\{-\Omega_{lib}^{\,2}\big/2(\Omega_{n,lib})^2\right\},\tag{8}$$

where $\Omega_{n,lib}$ is the parameter (width) of the distribution function.

Next we analyze a few double-pulse excitation scenarios to manipulate the optical vibrational responses. We have mentioned above that the laser control of the induced anisotropy in liquid bases on the constructive or destructive interference of the vibrational packets. It leads to the enhancement or the suppression of the vibrational responses in the total OKE signal. We consider the enhancement or the suppression of the vibrational mode amplitude after impact of the second pump pulse in relation to the amplitude of the mode response without the impact of the given pulse. The modeling of the vibrational responses with the multiple-pulse excitation (see (1) – (3) and (5)) takes into account the pulse length, the pulse intensities and the relaxation processes. Let us consider that the train of n pulses impacting on the system is characterized by the intensities I_p^i and the delays with respect to the first pulse τ_{1i}. For the qualitative analysis we assume the pulse length is considerably less than the period of the intramolecular vibrations. This enables one to simplify the formula (5) by the assumption that the excitation pulses have δ-function form:

$$I_p(t) = \sum_{i=1}^{n} I_p^i \delta(t - \tau_{1i})$$

$$r_{osc}(t) = (\alpha_{osc})^2 (\Omega'_{osc})^{-1} \sum_{i=1}^{n} H(\tau_{1i}) I_p^i \exp\left(-(t - \tau_{1i})/\tau_{osc}\right) \sin\left(\Omega'_{osc}(t - \tau_{1i})\right)$$

$$\Omega'_{osc} = \left(\Omega_{osc}^2 - \tau_{osc}^{-2}\right)^{1/2}\tag{9}$$

where $H(\tau)$ is Heaviside function. It is possible to adjust the intensity of the second pulse so the amplitude of the vibrations excited by the second pulse equals to the amplitude of the vibrations excited by the first pulse $I_p^2 = I_p^1 \exp\left(-\tau_{12}/\tau_{osc}\right)$. Then the vibrational response after the impact of the second pulse is

$$r_{osc}(t > \tau_{12}) \propto 2(\alpha_{osc})^2(\Omega'_{osc})^{-1}I_p^1 \exp(-t/\tau_{osc})\cos(\tfrac{1}{2}\Omega'_{osc}\tau_{12})\sin(\Omega'_{osc}(t - \tfrac{1}{2}\tau))$$

$$\tag{10}$$

.

To estimate the change of the amplitude A_{osc} of the response r_{osc} after the impact of the second pulse $(\tau > \tau_{12})$ with respect to the amplitude before the impact of the second pulse $(\tau < \tau_{12})$ we use the expression:

$$A_{osc}(\tau > \tau_{12})/A_{osc}(\tau < \tau_{12}) = 2\left|\cos(\tfrac{1}{2}\Omega'_{osc}\tau_{12})\right|. \tag{11}$$

Thus the expression (11) shows that the second pulse will increase the amplitude of the vibrational mode Ω'_{osc} in the OKE signal at the delay $\tau_{12} = 2\pi k/\Omega'_{osc}$ where $k \in N$. When the delay τ_{12} and the intensities of the pulses satisfy the conditions

$$\begin{cases} \tau_{12} = (\pi + 2\pi k)/\Omega'_{osc}, \\ I_p^{(2)} = I_p^{(1)}\exp(-\tau_{12}/\tau_{osc}) \end{cases} \tag{12}$$

the second pulse will remove completely the response of the mode Ω'_{osc} from the OKE signal. We propose to employ the double-pulse excitation for the selection spectroscopy of the vibrational and rotational responses in liquid.

Next we consider some of the double-pulse excitation scenarios. Let us assume that the medium have only one Raman-active vibrational mode Ω'_{osc}. So the OKE signal is the superposition of the vibrational and rotational responses. In this case the double-pulse excitation selects the rotational responses by implementation of the condition (12) with minimal delay τ_{12}. And the OKE signal will include the rotational responses only in the region $(\tau > \tau_{12})$ after suppression of the response of the mode Ω'_{osc}. In according to (6) and (7) the amplitude of the rotational responses increases proportional to the intensity of the second pulse.

The condition (12) shows that the vibrational response can be suppressed at the different values of the delay $\tau_{12} = (\pi + 2\pi k)/\Omega'_{osc}$. Since the completely suppression of the vibrational response requires the condition $I_p^2 = I_p^1 \exp\left(-\tau_{12}/\tau_{osc}\right)$ the relation of the second pulse intensity and the delay τ_{12} is associated with the relaxation of the intramolecular vibration. Thus the relation makes it possible to obtain the value of the relaxation constant τ_{osc} directly from the experiment without the modeling of the total OKE signal.

If the molecules of liquid have two Raman-active modes $\Omega'^{(1)}_{osc}$ and $\Omega'^{(2)}_{osc}$ the double-pulse excitation under the condition (12) will select the response of any mode in the total OKE signal. In addition there are some scenarios depending on specified delay τ_{12} so the second pulse suppresses the amplitude of one mode and at the same time enhances the amplitude of another mode in the total OKE signal.

It is impotent point to be made that a combined analysis of different scenarios improves the accuracy of decomposition of the total OKE signal into the molecular responses.

EXPERIMENTAL

Below we consider an application of some double-pulse scenarios to realize the selective spectroscopy of the molecular motions in neat liquids of carbon tetrachloride CCl_4 and chloroform $CHCl_3$. The liquids are transparent for visual light and in the area of the laser carrier frequency (790 nm). The molecules have the Raman-active low-frequency intramolecular modes exciting by 30 fs laser pulses. The symmetry of CCl_4 molecule is the reason for low amplitude of the rotational responses relative to the amplitude of the vibrational responses. The OKE signal of $CHCl_3$ liquid contains high amplitude vibrational and orientational responses.

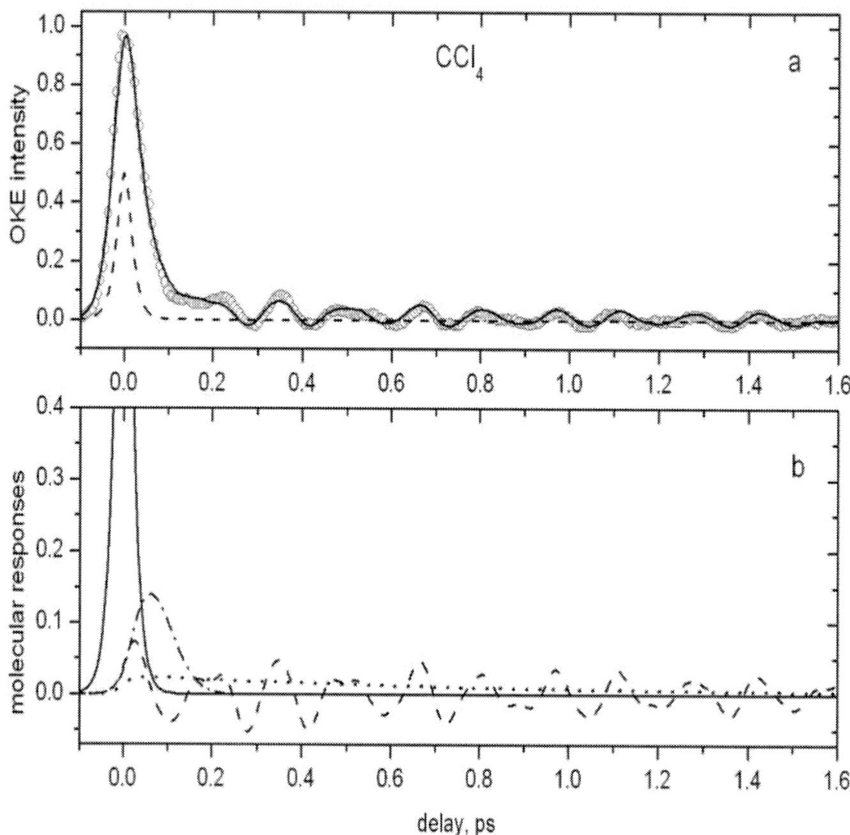

Figure 1. The single-pulse excitation OKE signal in CCl_4 liquid with $\tau_{pul} = 40$ fs. (a) -o- the experimental data, solid line – the simulation, dashed line – the envelope of the pulse intensity. (b) solid line – the electronic response R_{el}, dashed line – the vibrational response R_{vib}, dash-dotted line – the librational response R_{lib}, dotted line – the orientational response R_{or}.

The figure 1a shows the OKE signal of CCl_4 liquid at the single-pulse excitation. The components of the total OKE signal are shown on figure 1b. The figure 2 shows the imaginary component of the Fourier-transform of the OKE signal at the single-pulse excitation (figure 1a) and the Raman data. The figure 2 reveals that the 216 cm^{-1} and 312 cm^{-1} modes are excited efficiently. The 458 cm^{-1} mode is not detected in the OKE signal. The modeling of the experimental data of single- and double-pulse scenarios uses the same set of the calculation parameters: $\Omega_{osc}^{(1)} = 216 \pm 4$ cm^{-1}, $\tau_{osc}^{(1)} = 1.9 \pm 0.3$ ps, $\Omega_{osc}^{(2)} = 312 \pm 6$ cm^{-1}, $\tau_{osc}^{(2)} = 1.1 \pm 0.1$ ps, $\Omega_{osc}^{(3)} \equiv 458$ cm^{-1}, $\tau_{osc}^{(3)} \equiv 1.9$ ps, $\tau_{or} = 0.5 \pm 0.07$ ps, $\Omega_{n.lib} = 117 \pm 8$ cm^{-1}. We accept the value of $\Omega_{osc}^{(3)}$ is equal to the Raman value of the mode and the relaxation time $\tau_{osc}^{(3)} \equiv \tau_{osc}^{(1)}$. We are mainly interested in the responses of the coherent intramolecular vibrations r_{osc}^{i}. The modeling reveals that the laser pulses excite all vibrational modes. By the monitoring of the coherent vibrations we lost the high-frequency response due to the convolution of the molecular response $r_{osc}^{(3)}$ and the probe pulse (see (1) and (2)).

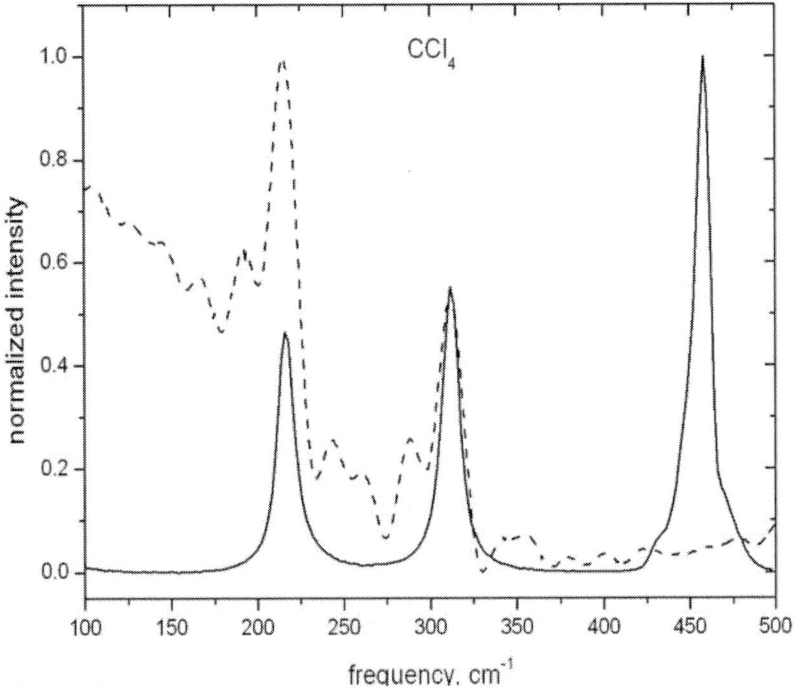

Figure 2. Solid line – the Raman data of CCl$_4$ liquid obtained on a standard DFS-52 Raman spectrometer, dashed line – Fourier-transform (imaginary component) of the OKE signal in CCl$_4$ liquid shown on figure 1a.

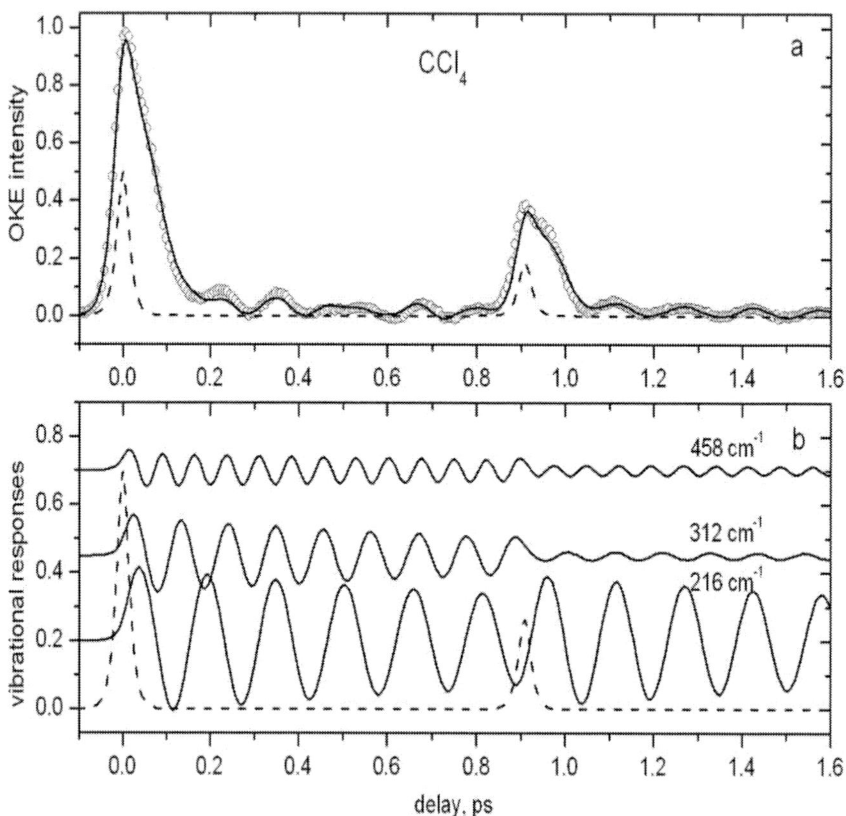

Figure 3. The double-pulse excitation of the OKE signal in CCl$_4$ liquid with $\tau_{pul} = 30$ fs, the delay between the pump pulses $\tau_{12} = 908$ fs and their intensity ratio $I_P^{(2)}/I_P^{(1)} = 0.38$.
(a) -○- the experimental data, solid line — the simulation, dashed line — the envelope of the double-pulse intensities. (b) solid lines — the vibrational responses of the 216 cm^{-1}, 312 cm^{-1} and 458 cm^{-1} modes, dashed line — the envelope of the double-pulse intensities.

Thus the response of the 458 cm^{-1} mode is not detected in the OKE signal of CCl$_4$ (see figure 2) we focus onto the manipulation of the amplitudes of the 216 cm^{-1} and 312 cm^{-1} modes. We analyze two scenarios of double-pulse excitation. (1) the second pulse enhances the amplitude of the 216 cm^{-1} mode and at the same time suppresses the amplitude of the 312 cm^{-1} mode.

Alternatively (2) the second pulse enhances the amplitude of the 312 cm^{-1} mode and suppresses the amplitude of the 216 cm^{-1} mode. The values of the relative intensities of the pump pulses and values of delays τ_{12} were adjusted by monitoring of the OKE signal in the experiments.

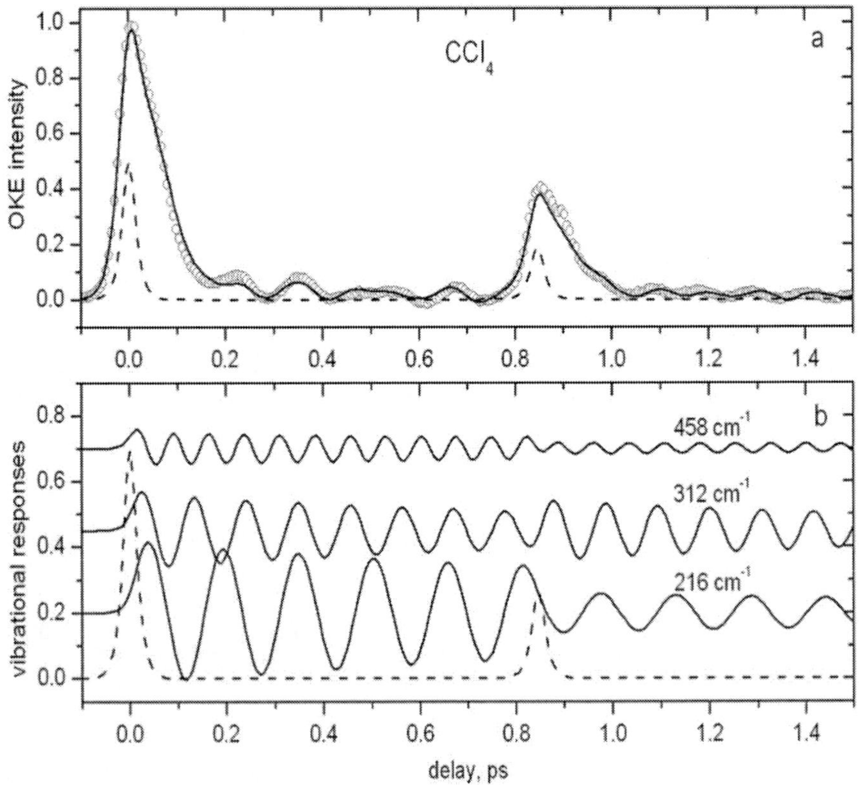

Figure 4. The double-pulse excitation of the OKE signal in CCl$_4$ liquid with $\tau_{pul} = 30$ fs, the delay between the pump pulses $\tau_{12} = 845$ fs and their intensity ratio $I_p^{(2)}/I_p^{(1)} = 0.38$.

(a) -o- the experimental data, solid line – the simulation, dashed line – the envelope of the double-pulse intensities. (b) solid lines – the vibrational responses of the 216 cm^{-1}, 312 cm^{-1} and 458 cm^{-1} modes, dashed line – the envelope of the double-pulse intensities.

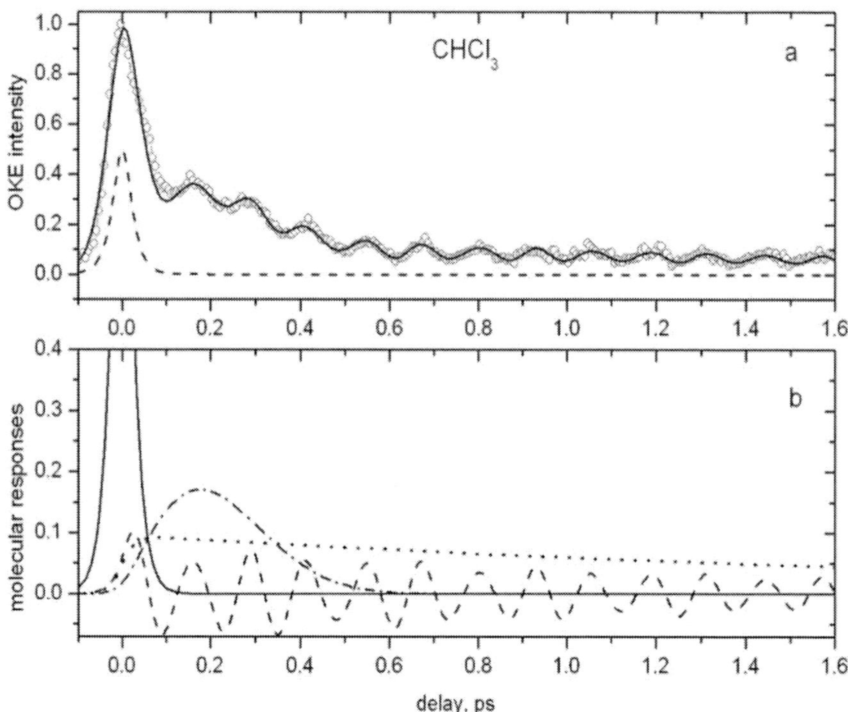

Figure 5. The single-pulse excitation of the OKE signal in $CHCl_3$ liquid with $\tau_{pul} = 40$ fs.

(a) -○- the experimental data, solid line – the simulation, dashed line – the envelope of the pulse intensity. (b) solid line – the electronic response R_{el}, dashed line – the vibrational response R_{vib}, dash-dotted line – the librational response R_{lib}, dotted line – the orientational response R_{or}.

The suppression of the 312 cm^{-1} mode response and the enhancement of the 216 cm^{-1} mode response by the double-pulse scenarios are shown on the figure 3. The scenario is specified by the delay $\tau_{12} = 908$ fs. The values of the vibration phases at the impact of the second pulse equal to 11.7π for the 216 cm^{-1} mode, 16.9π for the 312 cm^{-1} mode and 24.8π for the 458 cm^{-1}. The values of the expression (11) estimated the enhancement/suppression efficiency of the mode amplitudes equal to 1.8, 0.2 and 0.5, respectively.

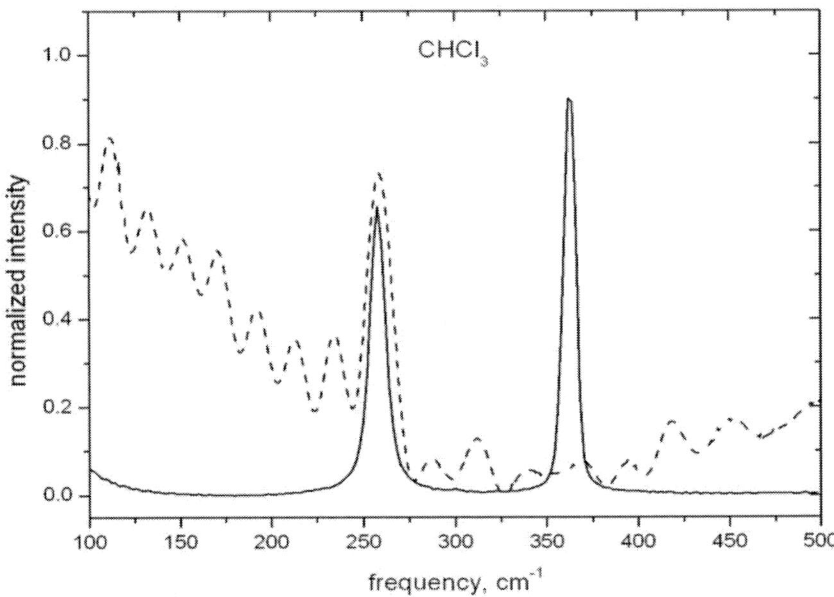

Figure 6. Solid line – the Raman data of CHCl$_3$ liquid obtained on a standard DFS-52 Raman spectrometer, dashed line – Fourier-transform (imaginary component) of the OKE signal in CHCl$_3$ liquid shown on figure 5a.

The alternative scenario uses double-pulse excitation to enhance the amplitude of the mode 312 см$^{-1}$ and to suppress the amplitude of the mode 216 см$^{-1}$ (figure 4). The value of the delay $\tau_{12} = 845$ fs, the values of the vibrational phases at the impact of the second pulse are 10.9π for the 216 cm^{-1} mode, 15.8π for the 312 cm^{-1} and 23.2π for 458 cm^{-1} mode. The values of the expression (11) for the modes are 0.2, 1.9 and 0.5, respectively.

The single-pulse excitation of the OKE signal in CHCl$_3$ liquid is shown on the figure 5a. The figure 5b shows the molecular components of the OKE signal. The modeling of the single-pulse and double-pulse scenarios shown on the figures 5, 7, 8 uses the same set of the calculation parameters: $\Omega_{osc}^{(1)} = 216 \pm 4$ cm^{-1}, $\tau_{osc}^{(1)} = 1.9 \pm 0.3$ ps, $\Omega_{osc}^{(2)} = 312 \pm 6$ cm^{-1}, $\tau_{osc}^{(2)} = 1.1 \pm 0.1$ ps, $\Omega_{osc}^{(3)} \equiv 458$ cm^{-1}, $\tau_{osc}^{(3)} \equiv 1.9$ ps, $\tau_{or} = 0.5 \pm 0.07$ ps, $\Omega_{n.lib} = 117 \pm 8$ cm^{1}. The figure 6 shows the imaginary part of Fourier-transform of the OKE signal on the figure 5a and the Raman data. The figure reveals the two Raman-active vibrational modes of 258 cm^{-1} и 362 cm^{-1} in the CHCl$_3$ molecules. But only one mode of 258 cm^{-1} having intensity response is identified well in the OKE signal and Fourier-transform data. So we do not obtain the accurate response of 362 cm^{-1} mode by analyzing the OKE signal at singe-pulse excitation.

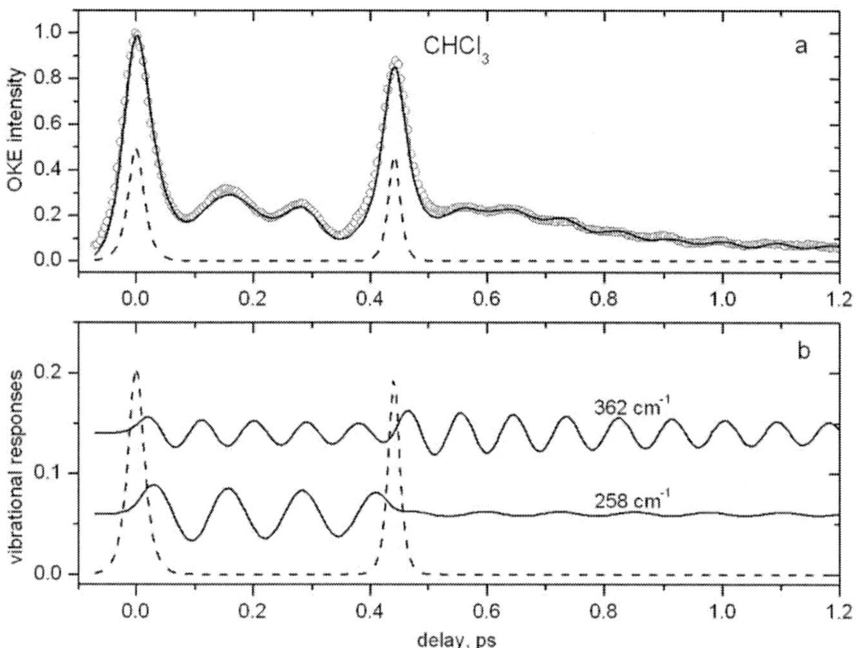

Figure 7. The double-pulse excitation of the OKE signal in CHCl$_3$ liquid with $\tau_{pul} = 30$ fs, the delay between the pump pulses $\tau_{12} = 440$ fs and their intensity ratio $I_p^{(2)}/I_p^{(1)} = 0.95$.

(a) -○- the experimental data, solid line – the simulation, dashed line – the envelope of the double-pulse intensities. (b) solid lines – the vibrational responses of the 258 cm^{-1}, and 362 cm^{-1} modes, dashed line – the envelope of the double-pulse intensities.

The figure 7 shows the double-pulse excitation scenario having the delay $\tau_{12} = 440$ fs. The phase values of vibrational modes at the impact of the second pulse are 7π for 258 cm^{-1} mode and 9.8π for 362 cm^{-1} mode. The values of the expression (11) for the modes are 0 and 1.95 respectively. In the scenario the second pulse suppresses the amplitude of the 258 cm^{-1} mode and enhances the amplitude of the 362 cm^{-1} mode. By this way we observe clearly the response of 362 cm^{-1} mode in the OKE signal as a result of the double-pulse excitation.

The figure 8 shows the suppression the of 258 cm^{-1} mode at the delay $\tau_{12} = 60$ fs. The phase values of vibrations at the impact of the second pulse are 0.9π for 258 cm^{-1} mode and 1.3π for 362 cm^{-1} mode. The values of the expression (11) for the modes are 0.2 and 0.9 correspondingly. We point out the second pulse manipulates both the vibrational and rotational responses. And the amplitude of the rotational responses enhances by the second pulse in all of scenarios. So the scenario of the double-pulse excitation on the figure 8 selects the rotational responses in the OKE signal.

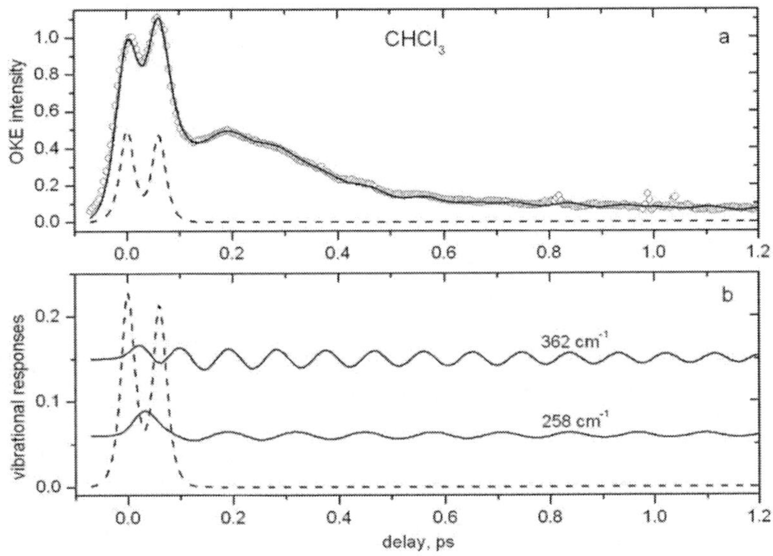

Figure 8. The double-pulse excitation of the OKE signal in CHCl$_3$ liquid with $\tau_{pul} = 35$ fs, the delay between the pump pulses $\tau_{12} = 60$ fs and their intensity ratio $I_p^{(2)}/I_p^{(1)} = 0.94$.

(a) -o- the experimental data, solid line – the simulation, dashed line – the envelope of the double-pulse intensities. (b) solid lines – the vibrational responses of the 258 cm^{-1}, and 362 cm^{-1} modes, dashed line – the envelope of the double-pulse intensities.

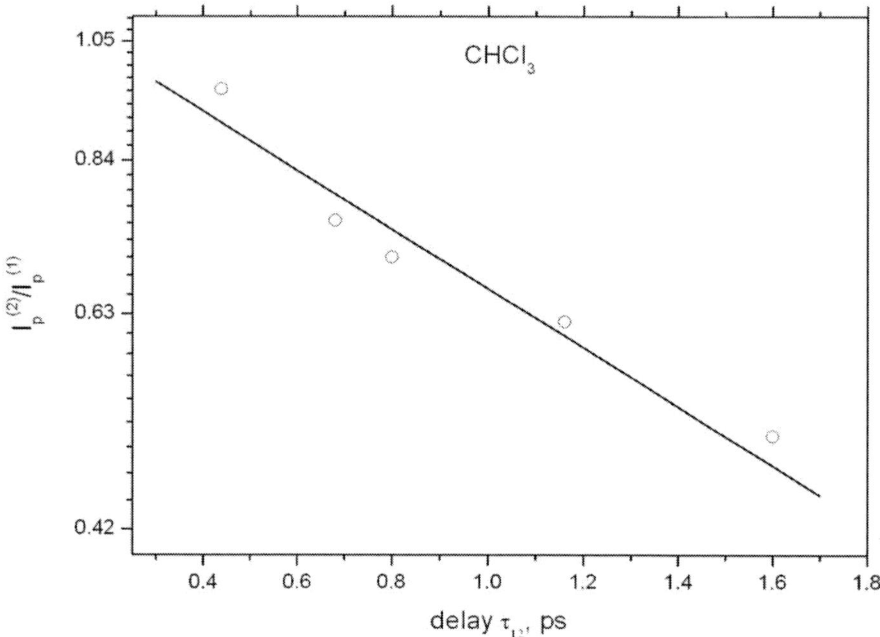

Figure 9. ○ - the relationship between the intensity ratio $I_p^{(2)}\big/I_p^{(1)}$ and the delays τ_{12} upon condition of the suppression of the 258 cm^{-1} mode. Solid line – the function exp(-τ_{12}/1.8).

Another important aspect is demonstrated on the figure 9, where the relationship between the intensity ratio $I_p^{(2)}\big/I_p^{(1)}$ and the delays τ_{12} upon condition of the suppression of the 258 cm^{-1} mode is shown. The analysis of the relationship based on the expression (12) obtains the values of the relaxation constant $\tau_{osc}^{(1)} = 1.8 \pm 0.2$ ps without the modeling the fifth-order optical response of the total OKE signal.

CONCLUSION

The scenarios considered above are the examples of the double-pulse non-resonant excitation for the manipulation of the optical molecular responses in the OKE signal. The constructive or destructive interferences of the coherent vibrational responses are the main factor of the enhancement or the suppression of the mode amplitudes in the OKE signal. The second pump pulse enhances the rotational responses at all cases of the double-pulse excitation. So the different scenarios of the coherent molecular dynamics are carried out by the variation of the delay τ_{12} and the relative intensities of the pump pulses $I_p^{(2)}/I_p^{(1)}$. Above we have considered some examples of the double-pulse scenarios to select the interested molecular responses in the total OKE signal. Especially we employ double-pulse excitation (1) to suppress the vibrational responses and enhance the orientational responses, (2) to suppress the amplitude of one vibrational mode and to enhance the amplitude of the another mode and thus to select the response of one mode in the total OKE signal, (3) to suppress the mode response at the different delay τ_{12} and in accordance with (12) to obtain the value of the relaxation constant τ_{osc}. Thus, it was shown that the time-resolved polarization selective spectroscopy of the molecular vibrational and rotational dynamics in liquid can be based on the double-pulse laser control.

Another important aspect of the double-pulse excitation compared with the single-pulse excitation is the greater number of the control parameters, while the number of the constants described medium is the same. It leads to the appreciable improvement in the procedure of the OKE signal decomposition into the molecular responses by modeling the different double-pulse excitation scenarios.

The work was supported by Russian Ministry of Science and Education (state contract No. 16.552.11.7008).

REFERENCES

[1]   G.Steinmeyer, J.Opt. *A :Pure Appl. Opt.* 5, R1 (2003).

[2]   A.H.Zewail, Femtochemistry: *Ultrafast Dynamics of the Chemical Bond* (Singapore: World Scientific Series in the 20-th Century Chemistry, 1994. p.912).

[3]   R.N.Zare, *Science* 279, 1875 (1998).

[4]   D.A.Wiersma, *Femtosecond Reaction Dynamics* (Amsterdam: Horth Holland, 1994.).

[5]   J.L.Herek, W.Wohlleben, R.J.Cogdell, D.Zeidler, M.Motztkus, *Nature* 417, 533 (2002).

[6]   V.S.Batista, *Science* 326, 245 (2009).

[7]   A.Lindinger, C.Lupulescu, M.Plewicki, F.Vetter, A.Merli, S.M.Weber, L.Wöste, *Phys.Rev.Lett.* 93, 033001 (2004).

[8]   R.J.Levis, G.M.Menkir, H.Rabitz, *Science* 292, 709 (2001).

[9]   V.I.Prokhorenko A.M.Nagy, S.A.Waschuk, L.S.Brown, R.R.Birge, R.J.D.Miller, *Science* 313, 1257 (2006).

[10]   A.M.Weiner, D.E.Leaird, G.P.Wiederrecht, K.A.Nelson, *J.Opt.Soc.Am.* B 8, 1264 (1991).

[11]   T.Dekorsy, W.Kutt, T.Pfeifer, H.Kurz, *Europhys.Lett.* 23, 223 (1993).

[12]   H.Hase, K.Mizoguchi, H.Harima, S.Nakashima, M.Tani, K.Sakai, M.Hangyo, *Appl.Phys.Lett.* 69, 2474 (1996).

[13]   M.F.DeCamp, D.A.Reis, P.H.Bucksbaum, R.Merlin, *Phys.Rev.*B 64, 092301 (2001).

[14]   M.Hase, M.Kitajima, S.Nakashima, K.Mizoguchi, *Appl.Phys.Lett.* 83, 4921 (2003).

[15]   C.A.D.Roeser, M.Kandyla, A.Mendioroz, E.Mazur, *Phys.Rev.*B 70, 212302 (2004).

[16]   A.Q.Wu, X.Xu, *Appl.Phys.Lett.* 90, 251111 (2007).

[17]   O.V.Misochko, M.V.Lebedev, H.Schäfer, T.Dekorsy, *J.Phys.Condens. Matter* 19, 406220 (2007).

[18]   C.M.Liebig, Y.Wang, X.Xu, *Optics Express* 18, 20498 (2010).

[19]   R.Ruhman, L.R.Williams, A.G.Joly, B.Kohler, K.A.Nelson, *J. Phys. Chem.* 91, 2237 (1987).

[20]   D.McMorrow, W.T.Lotshaw, G.A.Kenney-Wallace, *IEEE J. Quantum Electronics* 24, 443 (1988).

[21]   R.Righin, *Science* 262, 1389 (1993).

[22]   R.G.Brewer, C.H.Lee, *Phys.Rev.Lett.* 21, 257 (1968).

[23]   R.Cubedduye, R.Polloni, C.A.Sacchi, O.Svelto, *Phys.Rev.*A 2, 1955 (1970).

[24]   T.Kobayashi, A.Terasaki, T.Hattori, K.Kurokawa, *Appt.Phys.*B 47, 107 (1988).

[25]   D.McMorrow, *Opt.Comm.* 86, 236 (1991).

[26]   Y.Tanimura, S.Mukamel, *J.Chem.Phys.* 99, 9496 (1993).

[27]   T.Steffen, J.T.Fourkas, K.Duppen, *J.Chem.Phys.* 105, 7364 (1996).

[28]   S.A.Moiseev, V.G.Nikiforov, *Quantum Electron* 34, 1077, (2004).

[29]   V.G.Nikiforov, V.S.Lobkov, *Quantum Electron* 36, 984, (2006).

In: Molecular Dynamics
Editors: D. E. Garcia and P. J. Green

ISBN: 978-1-62081-545-8
© 2012 Nova Science Publishers, Inc.

Chapter 4

ZnO NANO-STRUCTURES FOR BIOSENSING APPLICATIONS: MOLECULAR DYNAMICS SIMULATIONS[*]

Safaa Al-Hilli[†] and Magnus Willander
Department of Science and Technology,
Campus Norrköping, Linköping University, Sweden

ABSTRACT

ZnO nanostructure is a material that is central for many nano-technology applications, such as chemical and biological sensors. A systematic molecular dynamics study for the behavior of water droplet and electrolyte solutions interacting with ZnO were done. The contact angle of a water droplet on ZnO polar slabs and nanorods/-tubes array changes significantly as a function of the ZnO-water interaction energy and nanostructure geometry. The water contact angle served as a criterion to tune the intermolecular interactions.

To recover a hydrophilic surface, voltage range from 1 to 20 volt were applied along the z-axis of the system to simulate the electrowetting behavior case. ZnO nanotube was used to study the permeation of water

---

[*] A version of this chapter also appears in *Advances in Nanotechnology, Volume 4*, edited by Zacharie Bartul and Jerome Trenor, published by Nova Science Publishers, Inc. It was submitted for appropriate modifications in an effort to encourage wider dissemination of research.
[†] E-mail: Safaa.Al-Hilli@itn.liu.se

for equilibrium and applied voltage cases, illustrating the influence of the surface topography and the intermolecular parameters and surface charges on permeation kinetics. We also studied the ionic currents through ZnO nanotubes simulating the case as field effect transistor (FET) of NaCl, KCl, CaCl2, and MgCl2 electrolyte solution for different concentrations (0.1, 0.5, 1.0, 5.0, and 10.0M) and calculate the solution conductance of these ions by applying a voltage difference (1-20V) along the z-axis of the system. We achieved by these molecular dynamics simulations the characteristic behaviors of ZnO nanostructure-electrolyte solution interactions and how it is suitable to use as ion selective sensor for intracellular microenvironment.

1. INTRODUCTION

The role of intracellular ions in signaling concentrations and pathways have followed technological breakthroughs in biochemistry, genetics, and developmental biology. Yet the fundamental mechanisms by which the total intracellular ionic state influences the cell response to its chemical and mechanical environment are still undetermined. This deficiency is due in part to limitations of analytical methods for assessing intracellular ion flux in living cells. Many measurements methods rely on the addition of chemical reagents such as ion binding dyes that may alter cell metabolism and provide only limited spatial resolution. Other deficiencies in sensing technologies can be attributed to the difference in size between the molecular processes taking place and the sensor. Recent advances in nanofabrication allow for the creation of new type of sensors. These probes introduce a new methodology controlled insertion into the interior environment of living cells. ZnO nanostructures have recently attracted considerable attention for the detection of chemical ions and biological molecules [1-7]. Among a variety of nano-sensor systems, the ZnO nanostructure electrochemical probe is one that offers high sensitivity and real-time detection. In this work we will try to simulate a probe from a single ZnO nano-rod/-tube or array which capable of electrochemically characterizing the simulated cell interior solutions. The use of a single ZnO nanotube probe will allow for sensing of concentrations and fluxes of electrochemically active ions. A detailed theoretical study of these interactions will be performed using molecular dynamics simulations showing spontaneous and continuous filling of a ZnO nanotube with ordered chains of water molecules. We will present our results as case study for ZnO hexagonal polar slabs interactions of wetting and electrowetting, wetting and electrowetting of ZnO nano-rods/-tubes array, and transport properties of water (filling) and ionic current of (Cl^-, K^+, Na^+, Mg^{2+}, and Ca^{2+}) ions confined in nanoscale one-dimensional channels multi-wall ZnO nanotube which have great interests for physics, biology and material science applications.

2. CASES STUDY

The interface of a solid in contact with liquid is of fundamental importance for a wide variety of applications, particularly biosensors. When hydrophobic solids are immersed in liquid, this process is called wetting. The controls of the wetting properties of the sensing surface are important for enhancing the sensitivity and reducing the liquid consumption. One method for modifying the hydrophobicity of a surface is through the application of an external electric field. Such a phenomenon is referred to as electrowetting, where the surface experiences a change in wettability upon the application of an electric field, which initiates a change in the contact angle between liquid-solid systems. There are many factors which affect the performance of an electrowettable surface, such as surface energies and tensions of the materials. The control of the contact angle has been used to drive liquid droplets in micro- or nanosize channels on biochemical and environmental sensing devices. In such applications, a faster contact angle transition rate is desired for prompt control of liquid movement. Limited work has been done to understand how different surface structures and morphologies affect the surface wettability of ZnO nanostructures.

2.1. ZnO Hexagonal Polar Surfaces Slab-Water Interaction (Wetting and Electrowetting)

When a liquid is in contact with an inert solid phase, the liquid "wets" the surface. Liquid molecules at the solid-liquid interface are now in a different environment than the ones that are either in the bulk or at the exposed surface. Those molecules feel two kinds of forces: cohesive forces acting between like molecules and adhesive forces acting between different molecules. The balance between cohesive and adhesive forces determines the wetting properties of the surface. When the cohesive forces of water can be counterbalanced by the adhesive forces of the substrate, a liquid droplet tends to spreads over the surface. When the cohesive forces of water are stronger than the adhesive forces of the substrate, the droplet tries to avoid the surface, keeping its spherical shape and reducing the surface tension. For a perfect homogeneous, horizontal, isotropic solid surface the contact angle of a liquid droplet (sufficiently large so that the effects of the three-phase contact line can be neglected) is obtained by the Young's equation by balancing the horizontal component of the forces acting on the three-phase (solid-liquid-vapor) contact line [8]:

$$\gamma_{sv} = \gamma_{sl} + \gamma_{lv}\cos\theta_Y \tag{1}$$

$$\cos\theta_Y = \frac{\gamma_{sv} - \gamma_{sl}}{\gamma_{lv}} \tag{2}$$

where θ_Y is Young contact angle and γ_{ij} is the surface tension or surface free energy, respectively. The indices stand for the different interfaces: solid-vapor (sv), solid-liquid (sl), and liquid-vapor (lv). In the present context, when characterizing the interactions of solid surfaces with water, we refer to surfaces with $\theta_Y > 90^o$ as hydrophobic and those with $\theta_Y < 90^o$ as hydrophilic.

Young's equation is strictly valid only for macroscopic droplets at mechanical equilibrium with any adsorbed film of thickness l (see Figure 1) on a molecularly smooth horizontal solid surface.

For microscopic droplets the contact angle will be influenced by surface interactions and the nature of the three-phase solid-liquid-vapor contact line of length L_{slv}, which will contribute an additional free energy per unit length or line tension τ to the excess free energy of the droplet. This effect adds an excess energy term $\tau\, dL_{slv}$ to Eq. 1. Hence for a spherical droplet of contact angle θ and lateral radius r_B, we obtain the modified Young's equation [9]:

$$\gamma_{sv} = \gamma_{sl} + \gamma_{lv}\cos\theta_Y + \frac{\tau}{r_B} \tag{3}$$

which can alternatively be written as

$$\cos \theta_w = \cos \theta_Y - \frac{\tau}{r_B \, \gamma_{lv}}$$

(4)

where θ_Y is the contact angle in the limit of very large droplets $r_B \to \infty$. Hence from Eq. 4 the quantities τ and $\cos\theta_Y$ can be determined from a study of $\cos\theta_w$ as a function of $1/r_B$.

The contact angle θ_Y provides important information about the wettability of a surface. A liquid partially wets a surface if a large droplet possesses a finite contact angle, $\theta_Y > 0$. If, however, this liquid has a contact angle $\theta_Y = 0$ the liquid completely wets the surface and the surface is covered by a thick liquid film.

In electrowetting, generally we are dealing with droplets of partially wetting liquids on planar solid surface substrates where in most interested applications, the droplets are aqueous salt solutions with a typical size of the order of 1mm or less. In case of applying external electric field the electrowetting for homogeneous planar solid substrates can be described by Lippmann's equation of electrowetting which is based on general Gibbsian interfacial thermodynamics (see Figure 2) [10].

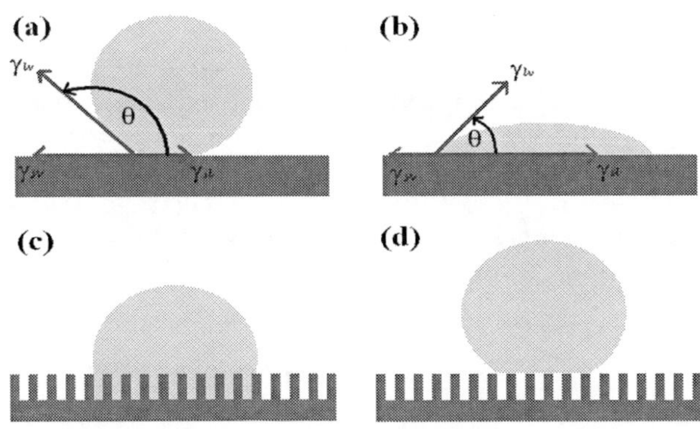

Figure 1. A small droplet in equilibrium over a horizontal surface with different degree of wetting; (a) the wetting is low and the contact angle is large, (b) the wetting is high and the contact angle is small. Droplets wetting rough surfaces; (c) Wenzel state with enhanced solid-liquid interfacial area, (d) the Cassie-Baxter state with entrapped air underneath the droplet (non-wetting condition).

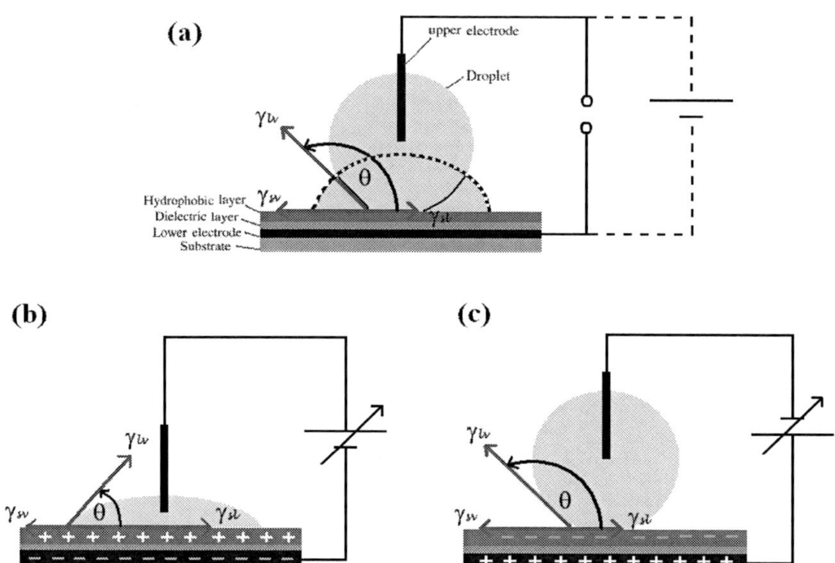

Figure 2. Electrowetting set-up; (a) partially wetting liquid droplet at zero voltage (solid line) and at high voltage (dash line). Operation of voltage between the droplet and the electrode (b) forward biasing voltage which changes the distribution of charges due to dielectric insulator and decreases the contact angle , (c) backward biasing voltage which reverse the surface polarity (hydrophobic surface) and increases the contact angle.

Surface wettability can be enhanced by application of electric voltage driving dipolar water molecules to the field-exposed region. Upon applying a voltage dV, an electric double layer builds up spontaneously at the solid-liquid interface consisting of charges on the metal surface on the one hand and a cloud of oppositely charged counter-ions on the liquid side of the interface. Since the accumulation is a spontaneous process, the effective interfacial tension of solid-liquid can be defined as:

$$d\gamma_{sl}^{eff} = -\sigma_{sl}dV$$

(5)

where $\sigma_{sl} = \sigma_{sl}(V)$ is the surface charge density of the counter ions. The voltage dependence of is calculated by integrating the above equation. In general, this integral requires additional knowledge about the voltage dependent γ_{sl}^{eff} distribution of the counter ions near the interface which is calculated on the basis of the Poisson-Boltzmann distribution. For the simplicity we assume that the counter-ions are all located at a fixed distance (in order of a few nm) d_H from the surface (Helmholtz model). In this case, the double layer has a fixed capacitance per unit area,

$$C_H = \frac{\varepsilon_o \varepsilon_l}{d_H}$$

(6)

where ε_l is the dielectric constant of the liquid.

We obtain[11]:

$$\gamma_{sl}^{eff}(V) = \gamma_{sl} - \frac{\varepsilon_o \varepsilon_l}{2d_H}(V - V_{pzc})$$

(7)

where is the potential difference of zero charge (metal surfaces acquire a spontaneous charge when immersed into electrolyte solution at zero voltage and V_{pzc} is the voltage required to compensate for this spontaneous charging), and γ_{sl} the chemical contribution to the interfacial energy is assumed to be independent of the applied voltage.

To obtain the response of the contact angle for an electrolyte droplet placed directly on an electrode surface we used this equation [12]:

$$\cos \theta_w = \cos \theta_Y + \frac{\varepsilon_o \varepsilon_l}{2 d_H \gamma_{lv}}(V - V_{pzc})$$

(8)

For typical values of $d_H = 2\,nm$, $\varepsilon_l = 81$, and $\gamma_{lv} = 72\,dyne/cm$, we find that the ration on the term is on the order of $1V^{-2}$. The contact angle thus decreases rapidly upon the application of a voltage.

To include the effect of the nanometer size of the water droplet on the WCA the Eq. 8 will be as follow:

$$\cos \theta_w = \cos \theta_Y - \frac{\tau}{r_B \gamma_{lv}} + \frac{\varepsilon_o \varepsilon_l}{2 d_H \gamma_{lv}}(V - V_{pzc})$$

(9)

When semiconductor material get in contact with liquid depletion will create which insulates the droplet from the electrode and the electric double layer builds up at the depletion-droplet interface. Since the depletion layer thickness d is usually much larger than d_H ,, the total capacitance of the system is reduced tremendously. The system can be described as two capacitors in series, namely the Helmholtz capacitor C_H and the semiconductor layer (depletion layer) C_d defined as:

$$C_d = \frac{\varepsilon_o \, \varepsilon_d}{d}$$

$$(10)$$

Since $C_d << C_H$, the total capacitance per unit area will be $C \approx C_d$. With this approximation, we neglect the finite penetration of the electric field into the liquid. As a result, we find that the voltage drop occurs within the depletion layer and the Eq. 7 is replaced by [11,12]:

$$\gamma_{sl}^{eff}(V) = \gamma_{sl} - \frac{\varepsilon_o \, \varepsilon_l}{2\,d} V^2$$

$$(11)$$

Here we assume that the surface of the depletion layer does not give rise to spontaneous adsorption of charge in the absence of an applied voltage, i.e., we set $V_{pzc} = 0$. We can rewrite the above equation in term of contact angle as:

$$\cos \theta_w = \cos \theta_Y + \frac{\varepsilon_o \, \varepsilon_l}{2\,d\,\gamma_{lv}} V^2$$

$$(12)$$

Then we add the nano-droplet modified term the above equation will be:

$$\cos \theta_w = \cos \theta_Y - \frac{\tau}{r_B \, \gamma_{lv}} + \frac{\varepsilon_o \, \varepsilon_l}{2\,d\,\gamma_{lv}} V^2$$

$$(13)$$

2.2. ZnO Nanorods or Tubes Array-Water Interaction (Wetting and Electrowetting)

There have been a lot of interests in studying the wetting behaviors of ZnO nano-rods/-tubes array, which is a technologically important for many promising applications in the fields of chemical and biological sensing where the surface wettability plays a very important role. These nanostructures were generally used to enhance the surface wettability of ZnO films and, in some cases, to obtain superhydrophobicity surfaces.

The wetting properties of the nanostructures are different from the smooth surfaces. The principles of wetting rough surfaces have been investigated by Wenzel [13] with the assumption that the entire surface under the drop is covered with liquid (see Figure 1c). The Young's equation is modified as:

$$\cos \theta_w = \alpha \frac{\gamma_{sv} - \gamma_{sl}}{\gamma_{lv}} = \alpha \cos \theta_Y \tag{14}$$

where θ_w is the contact angle predicted by Wenzel's model, θ_Y is the Young contact angle and α is the roughness factor, defined as the ratio of the actual area of a rough surface to the geometric projected areas.

For a rough hydrophobic surface however liquid may not completely penetrate into surface hollows and trapped air inside, forming a composite solid-air-liquid interface. Cassie and Baxter extended Wenzel's equation to describe the contact angle on such a surface (see Figure 1d) [14]:

$$\cos\theta_w = f_1 \cos\theta_1 + f_2 \cos\theta_2 \tag{15}$$

Where f_1 and f_2 are the area fractions of liquid-solid interface and liquid-air interface, respectively. As $f_1 + f_2 = 1$ and $\theta_2 = 180^o$ due to the non-wetting condition with air, the equation above can be reduced to:

$$\cos\theta_w = f_1 \cos\theta_1 - f_2 \tag{16}$$

$$\cos\theta_w = f_1(\cos\theta_y + 1) - 1$$

$$\tag{17}$$

This is in agreement with the fact that surface hydrophobicity improves when there is more air trapped between the liquid and solid surfaces.

We can idealize the ZnO nanorods as flat-top, hexagonal sectioned rods with a height of h and a top radius of the rod of R, as illustrated in Fig. 3. When a water droplet contacts the surface, the area fraction of liquid-solid interface f_1 can be determined by [15]:

$$f_1 = \frac{\left(\frac{3\sqrt{3}}{2}R^2\right) + 6R\,h'}{\frac{A}{n} + 6R\,h}$$

$$\tag{18}$$

where A is the projected area of the water droplet on the ZnO nanorod film surface, n is the number of nanorods in the area A $(A/n \geq (3\sqrt{3}/2)R^2)$, and h' $(h' \leq h)$ is the depth that water intrudes into the gaps between adjacent rods. When adjacent nanorods are closer (n increases) in a certain area A, the depth of intruding water h' will decrease. The decrease of h' will diminish the increase of f_1 caused by the increase of n and may even lead to a net decrease of f_1.

For a rough hydrophobic surface however we can use the Cassie and Baxter extended equation to describe the contact angle on such a surface in case of electrowetting and to include the effect of the nanometer size of the water droplet on the WCA, we modify Eq. 9 to become:

$$\cos\theta_w = \left[f_1(\cos\theta_Y + 1) - 1\right] - \frac{\tau}{r_B\,\gamma_{lv}} + \frac{\varepsilon_o\,\varepsilon_l}{2\,d_H\,\gamma_{lv}}\left(V - V_{pzc}\right)^2$$

$$\tag{19}$$

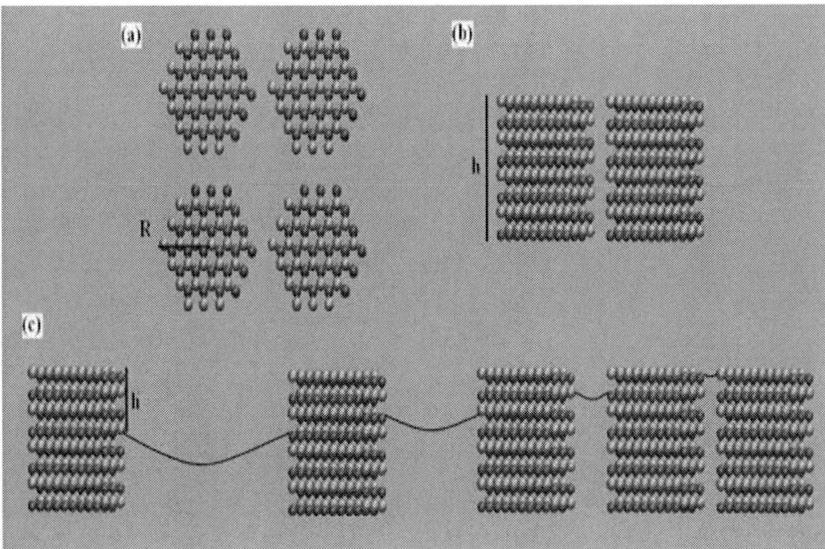

Figure 3. Schematic diagram of roughness factor calculation; (a) and (b) ZnO nanorods array representation which composite of a certain number of ZnO rods n of radius R and length h in area A. (c) Depth of the intruding water h' will decrease with increasing n (adjacent nanorods are closer).

We can call the extended equation as Lippmann-Cassie and Baxter electrowetting equation for nano-structure surfaces.

For the case where $V_{pzc} = 0$, the final equation will be:

$$\cos\theta_w = \left[f_1\left(\cos\theta_Y + 1\right) - 1\right] - \frac{\tau}{r_B\,\gamma_{lv}} + \frac{\varepsilon_o\,\varepsilon_l}{2\,d_H\,\gamma_{lv}}V^2 \qquad (20)$$

2.3 Water Permeation through ZnO Nanotube

Transport of water into and out of nanotube will be studied in this case. The geometry of a tube (its inner radius, length and shape) and the chemical character of the tube inner wall have a great influence on the permeation of water (see Figure 4).

We have adopted the Oliver et al. [16] model for permeation of water and ions through nano-pores. This approach model is used to label the two phase states that described the water permeation exhibits in the simulation.

We either find liquid-filled tube or vapor-filled one. Due to the small tube inner volumes "vapor" typically refers to zero or one to water molecules in the cavity. Where the water density in the tube is used as an indicator of the phase state, when the water density $b(t)$ rises above 0.65 of the density of bulk water, b_o ($b_o = 1.0\,g\,cm^{-3}$ at $T = 300\,K$ and $P = 1\,bar$) the liquid state is assigned to the phase state at time t, or when $b(t)$ drops below $0.25\,b_o$ the vapor state is assigned.

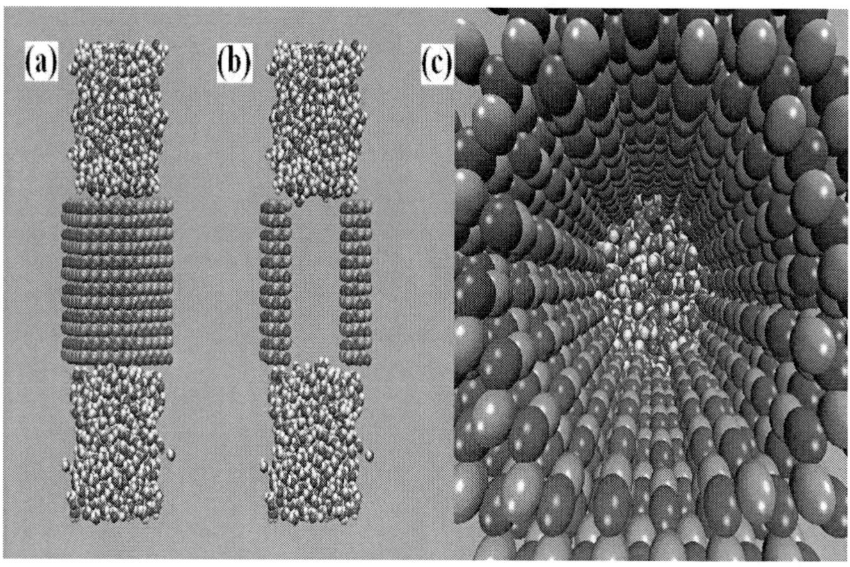

Figure 4. Permeation of water through a ZnO nanotube initial MD set-up; different representations of the system (a) side view, (b) y-axis cross section of the system, (c) inner view through the ZnO tube.

The model assumes the inner tube volume is $\pi r^2 L$, and this volume can exchange water molecules with the bulk water outside the tube, which acts as a particle reservoir at average chemical potential μ. The free energy that describe the tube in the closed and open state used as $\nabla U(T,V,\mu) = -PV$.

Then free energy difference between the vapor and the liquid state with the corresponding surface contributions will define as [16]:

$$\Delta U(r) = 2\left[\gamma_{lv} + \frac{1}{2}\Delta\mu\,\Delta b_{lv}\,L\right]\pi\,r^2 + 2\pi\,L\,\Delta\gamma_s\,r$$

(21)

where $\Delta\mu = \mu - \mu_{sat}$ is the distance of the state from saturation, $\Delta b_{vl} = b_l - b_v$ the difference in densities at saturation, and $\Delta\gamma_s = \gamma_{vs} - \gamma_{ls}$ the difference in surface free energies of the two phases with the wall. The term $\Delta\mu\,\Delta b_{vl}\,L$ is small for $L < 10\,nm$ as the system is close to phase coexistence (for $L \approx 1\,nm$ it is about 10^{-3} times smaller than $\gamma_{lv} = 17\,k_B\,T\,nm^{-2}$ when estimated from $\Delta b_{vl}\,\Delta\mu\,L \approx \Delta P\,L \approx 1\,bar \times L = 2.4 \times 10^{-3}\,k_B\,T\,nm^{-2} \times L,\;at\;T = 300\,K$ and will be neglected.

Only the difference between the surface free energies enters the model so we express it as the contact angle θ_Y, using the macroscopic definition from Young's equation [17]. Then Eq. 21 becomes [16]:

$$\Delta U(r, L, \theta_Y) = 2\pi\,r\gamma_{lv}\left(r + L\cos\theta_Y\right)$$

(22)

For fixed tube length L and a given tube material, characterized by θ_Y, the graph of $U(r)$ over the tube inner radius r describes a parabola containing the origin [18].

The connection between the model free energy $U(r)$ and the behavior of the system as observed in MD simulations, it is assumed that g_i as numbers of equilibrium discrete states (MD equilibrium trajectory), i.e. the openness $\langle g(r)\rangle$, is defined as [16]

$$\langle g(r)\rangle = \frac{1}{1 + \exp\left(-\dfrac{1}{k_B T}\Delta U(r)\right)}$$

(23)

The macroscopic relation due to Young and Lippmann can describe electrocapillarity in a planar confinement as (see Figure 5) [19]:

$$\cos\theta_w = \cos\theta_Y - \frac{W_{el}(V)}{2\gamma_{lv}} = \cos\theta_Y + \frac{CV^2}{2\gamma_{lv}}$$

(24)

Here $W_{el}(V)$ is the change in electrostatic free energy per unit area, associated with surface spreading of the liquid, wetting both walls (hence the factor 1/2), V is the voltage across the interface, and θ_Y the contact angle in the absence of electric field. The form of W_{el} depends on system geometry and material properties but is generally presumed to be proportional to the areal electric capacitance of the interface, C, and the potential drop across the interface squared.

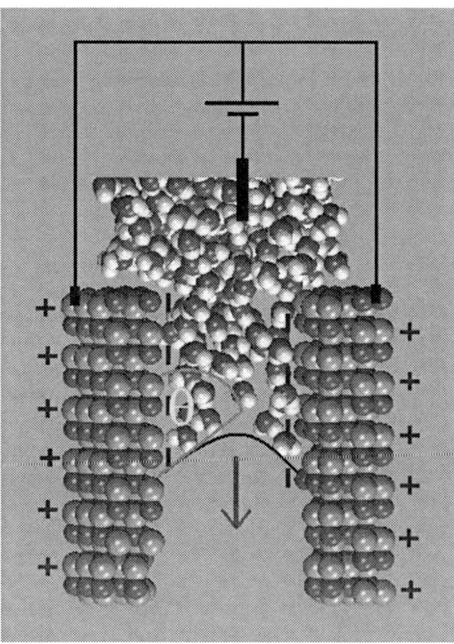

Figure 5. Water Electro-permeation through ZnO nanotube. In this case the ZnO tube walls act as a dielectric insulator.

2.4. Ionic Currents of Mg^{2+}, Ca^{2+}, K^+, and Na^+ Ions through ZnO Nanotube

To calculate the current-voltage characteristics of the electrolyte we will follow the mobility model based on molecular dynamics would generalize this approximation [20]. The potential variation through the nanotube with the following analytical expression, which has been shown to be valid from molecular dynamics in nanotubes [20]:

$$V(z) = \frac{V_0}{\pi} \tan^{-1}\left(\frac{z}{L_{eff}}\right)$$

(25)

where V_o is the external voltage across the device driving the ions through the nanotube and L_{eff} is a characteristic length (not the channel length) so that the potential achieves its electrode values at $z = \left[-L_z/2, L_z/2\right]$ where L_z is the channel length, with non-zero electric field at the electrodes. We assume the following approximation that reservoir resistance is negligible in comparison with nanotube resistance and the potential in the reservoir is constant.

Considering both anion density current J_a and cation density current J_c and neglecting the diffusion current, the current density for each ion type is given by [20,21]:

$$J_a = q\mu_a c_a \nabla\psi$$

(26)

$$J_c = q\mu_c c_c \nabla\psi$$

(27)

and the total current density is [20]:

$$J = J_a + J_c$$

(28)

where ψ is the electrostatic potential, c_a and c_c are the anion and cation concentration, μ_a and μ_c are the anion and cation mobility, respectively. Assuming no recombination inside the tube, we have [20]:

$$\nabla J_{a,c} = 0 \tag{29}$$

which by using the divergence theorem, implies that the current is constant through the nanotube. Due to the one dimensional nature of the external potential (Eq. 25), this condition (Eq. 29) is not fully satisfied. Therefore, we spatially average the current through the nanotube to eliminate the slight J-variations due to the slanted geometry of the nanotube [20]:

$$\langle I_{a,c} \rangle = \frac{1}{L} \int_{L/2}^{-L/2} dz \int \int_{S(z)} J_{a,c}(r) dS \tag{30}$$

Here L is the length of the tube and $S(z)$ is the nanotube cross section at ordinate z. In this context, we define each ion conductance in the nanotube as [20]:

$$G_{a,c} = \frac{\langle I_{a,c} \rangle}{V_o} \tag{31}$$

3. METHOD

To study the sensitivity of the contact angle to the ZnO-water interaction potential and the droplet size and shape, a series of MD simulations of water droplets on ZnO nanostructures are performed. The MD simulation techniques are described along with the details on how the contact angles, density profiles, and line tension are extracted from the simulations. In these cases, we will describe how to build atomic-scale models of ZnO hexagonal polar surface slabs and ZnO nano-rod/-tube arrays using computational procedures that mimic the surface wetting experiment. The shape and dimensions including diameter size for the water droplet were varied. The surface morphology and dimensions including diameter size, rod length, and density were varied. The surface wettability of ZnO films were examined by water contact angle measurements. Switchable wettability was also investigated on the both types of films (rods and tubes) by applying voltage difference across the system. Water permeation through ZnO nanotube was studied by varying the inner radius of the nanotube for different applied voltages. Ionic current through ZnO nanotube were studied by applying voltage along the z-axis of the system with varying the salt concentration of the electrolytes for NaCl, KCl, $CaCl_2$, and $MgCl_2$ solutions.

3.1. Molecular Dynamics

Molecular dynamics (MD) simulations can be used to complement the experimental data on the ZnO–water interaction. Since interactions with biomolecules in all biotechnical applications occur in an aqueous environment, an empirical force field for ZnO should be able to reproduce its surface wetting properties. For the ZnO-water dynamics we used the molecular dynamics program NAMD 2.8 [22] and the simulation outcomes were analyzed through our own routines programmed in Matlab 7.5 [23] and VMD 1.8.7 [24] programs. The universal force field (UFF) provided by Rappé et al. [25] were employed to produce hexagonal polar slab, nanorod and nanotube from ZnO. The UFF is a harmonic force field, in which the total potential energy expression is expressed as:

$$E = E_R + E_\theta + E_\phi + E_w + E_{vdW} + E_{el} \tag{32}$$

where E_R is the bond stretching energy, E_θ is the bond bending energy, E_ϕ is the torsions energy, E_w is the inversions energy, E_{vdW} is the ven der Waals interaction energy and E_{el} is the electrostatic interaction energy. The functional form of the above energy terms are given as follows:

$$E_R = k_1 (r - r_o)^2 \tag{33a}$$

$$E_\theta = k_2 (A_o + A_1 \cos\theta + A_2 \cos 2\theta) \text{, where} \tag{33b}$$

$$A_2 = \frac{1}{\left(2\sin^2\theta_o\right)},$$

$$A_1 = -4A_2\cos\theta_o,$$

$$A_o = A_2\left(2\cos^2\theta_o + 1\right),$$

$$E_\phi = k_3\left(1 \pm \cos n\phi\right) \tag{33c}$$

$$E_\omega = k_4\left[1 + \cos(n\chi - \chi_o)\right] \tag{33d}$$

$$E_{vdW} = D\left[\left(\frac{r^*}{r}\right)^{12} - 2\left(\frac{r^*}{r}\right)^{6}\right] \tag{33e}$$

$$E_{el} = \frac{q_i q_j}{\varepsilon\, r_{ij}} \tag{33f}$$

where k_1, k_2, k_3, and k_4 are force constants, θ_o is the natural bond angle, D is the van der Waals well depth, r^* is the van der Waals length, q_i is the net charge of an atom, ε is the dielectric constant, r_{ij} is the distance between two atoms. In the calculation, the difference of the non-polar and polar semiconductors is dependent of the net charge q_i being set to zero or not.

To parameterize a force field that can account for ZnO-water interactions we employed a potential energy function compatible with the CHARMM force field [26]. We began with previously existing UFF parameters for ZnO and refined them to reproduce the interactions of ZnO with water. The functional form used is:

$$E_{Total} = E_{Bond} + E_{Angle} + E_{vdW} + E_{Coulomb} \tag{34}$$

The first two terms on the right-hand side of Eq. 34 are harmonic potentials used to describe bond stretching and bending:

$$E_{bond} = \sum_{bonds\ i} k_i^{bond}\left(r_i - r_{0i}\right)^2 \tag{35}$$

$$E_{angle} = \sum_{angles\ i} k_i^{angle}\left(\theta_i - \theta_{oi}\right)^2 \tag{36}$$

Here the sums run over all bonds and bond angles; the parameters k_i^{Bond}, k_i^{Angle}, r_{oi}, and θ_{oi} describe the equilibrium values of the degrees of freedom. The bonded interactions of ZnO were taken from UFF but adjusted to fit Eqs. 35 and 36.

The last two terms in Eq. 34 describe the vdW and electrostatic nonbonded interactions that are the main focus of this work:

$$E_{vdW} = \sum_{i}\sum_{j>i} D_{ij}\left[\left(\frac{r_{ij}^{min}}{r_{ij}}\right)^{12} - 2\left(\frac{r_{ij}^{min}}{r_{ij}}\right)^{6}\right] \tag{37}$$

$$E_{Coulomb} = \sum_{i}\sum_{j>i} \frac{q_i q_j}{4\pi\varepsilon_o r_{ij}} \tag{38}$$

The vdW interactions are represented through a Lennard-Jones 6-12 potential, with two adjustable parameters: r_{ij}^{min}, the distance at which the energy between atoms i and j is minimal, and D_{ij}, the well depth. In the case of the oxygen atom, vdW parameters were taken from the CHARMM force field, assigning 3.5 Å to r_O^{min} and 0.15 kcal/mol to D_O. For zinc, r_{Zn}^{min} was set to 2.763 Å, and 0.25 kcal/mol to D_{Zn} corresponding to the zinc atomic radius [25]. The electrostatic interactions were calculated using Eq. 38, q_i and q_j being the partial atomic charges of atoms i and j, and ε_o being the vacuum permittivity. The surface charges corresponding to zinc q_{Zn} and oxygen q_O were tuned to reproduce the ZnO wettability. We employed the TIP3P model of water, since the CHARMM force field works best with this choice of model and its functional form conforms to Eq. 34.

3.2. Building ZnO Structures

3.2.1. ZnO Hexagonal Polar Slab

ZnO is polar semiconductor; his polarity came from bulk characteristic of hexagonal wurtzite materials that stems from the lack of a symmetry center and results in two inequivalent c-directions. The noncentrosymmetric structure and the partial ionicity of the bonds, which results from the differing Pauling electronegativity of the atoms, yield a net spontaneous polarization field along the c-axis. Therefore, the Zn-polar ZnO (0001) and O-polar ZnO (000$\bar{1}$) are characterized by different properties, such as morphology, dipole moment, surface charge, band bending, and chemical stability.

Most significantly, a strong polarization-induced surface charge of opposing sign characterizes the (0001) and the (000$\bar{1}$) polar surfaces. To generate the ZnO hexagonal polar slab structures we used the plugin of VMD program. The schematic procedure for building ZnO nanostructure is shown in Figure 6.

To create hexagonal polar slab of ZnO, we replicated the unit cell with lattice parameters a=3.249 Å and c=5.207 Å [27] for 26-26-2 times and yielding a system of 5408 atoms. This step was followed by the cut the slab to hexagonal shape having 4056 atoms with radius 42.237 Å and slab thickness 10.414 Å (see Figure 6d).

3.2.2. ZnO Nanorods

To generate the ZnO nanorods and nanotubes array structures we used the VMD program. The schematic procedure for building ZnO nanostructure is shown in Figure 6b. To build hexagonal rod of ZnO, we replicating the ZnO unit cell for 6-6-4 times and yielding a system of 576 atoms. Subsequently, the bulk system was cut in hexagonal shape to reproduce ZnO hexagonal rod crystal having 432 atoms with dimensions (r=9.747 Å and l_z=20.828 Å).

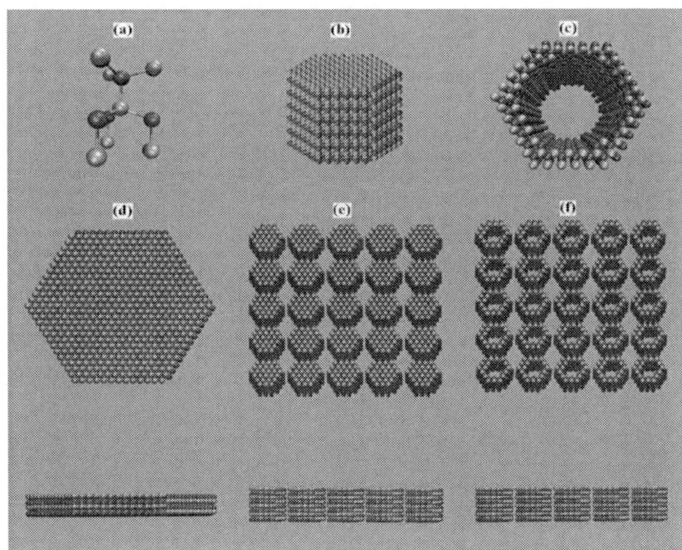

Figure 6. Building steps of ZnO nano-structures. (a) Unit cell of ZnO, which is replicated to generate the hexagonal shape of ZnO nanorod (b), the cylinder inner shape drilled to make the ZnO nanotube (c), a thin hexagonal slab of the ZnO polar surfaces were created by replicating the unit cell 26-26-2 times (d). ZnO nanorods/tubes array were created by replicating the ZnO rod/tube at equal distances in x- and y-axes, (e) for rods array and (f) for tubes array.

3.2.3. ZnO Nanotube

To model the ZnO nanotube, a hexagonal prism tube of ZnO, with length of 2.0828 nm, was constructed by removing atoms from the hexagonal ZnO nanorod structure; we produce a cylindrical tube of symmetric shape with inner radius 0.81225 nm. The length of the cylinder tubes equals the hexagonal ZnO nanotube length (see Figure 6c). However, the detailed shape of the cylinder tube surface is not fully regular and the presence of irregularities and deviations from a symmetrical cylinder structure are considered. These irregularities can affect the motion of water molecules or ions through the ZnO nanotube. The total charge of a nanotube was found to approximately be zero.

To make ZnO nanorods/tubes array we replicate the nanorod/tube structures at equal distance (in x and y directions) from each other's and we have varied these distance to change the density of nanorods/tubes in the array (in our case 2.2 nm center-center distance with rods or tubes density per area equal to 0.65 nm^{-2}) (see Figure 6e and 6f).

3.3. ZnO- Water Systems

3.3.1. Wetting

To simulate the water contact angle (WCA) of ZnO [28,29]; rectangular and spherical volumes of water were generated from a pre-equilibrated water box using the solvate plugin of VMD. A water box with side lengths 30-30-30 $Å^3$, and 40-38-30 $Å^3$ composed of 2478, and 4236 atoms and a water sphere with radius of 16 Å composed of 1428 atoms were placed on slab top of the ZnO hexagonal polar surfaces (see Figure 7a,b).

To perform MD simulations with full electrostatics under periodic boundary conditions, the total charge of the system has to be zero. To maintain electroneutrality of the systems for ZnO-water wetting interaction, the surface atoms were assumed as follows: two atoms were considered covalently bonded if they had a separation distance of 2 Å or less; oxygens with four bonds and zinc with four bonds were classified as a non-dangling type; oxygens with less than four bonds and zinc with less than four bonds were classified as a dangling oxygen type and dangling zinc type, respectively, or just as a dangling atom type.

All simulations involving ZnO-water systems were carried out using a CHARMM compatible form of the potential energy function (Eq. 34).

Simulations were performed with an integration time step of 1 fs for 1ps using the conjugate gradient method and then equilibrated for 2 ns in the *NVT* ensemble for periodic boundary conditions at 300 K, vdW interactions were calculated with a cutoff of 12 Å (switching function starting at 10 Å), and the long-range electrostatic forces were calculated using PME summation method. A Langevin thermostat was used to maintain a constant temperature in the NVT ensemble simulations. Simulations in the NpT ensemble were performed using a hybrid Nose-Hoover Langevin piston. To keep the ZnO structure rigid, the Zn and O atoms were either fixed or restrained to their original position by applying a harmonic force with a force constant of 10 kcal/mol/$Å^2$.

The water temperature remains stable during the MD runs; it has an average of 298.9 K and a standard deviation of 5.0 K. Figure 8 illustrates the initial and equilibrate configuration. Samples of the trajectory are stored every 0.1 ps.

3.3.2. Electrowetting

To simulate the electrowetting of the ZnO nanostructures, we applied a uniform electrical field to all atoms; it induces, at the beginning of the simulation, a rearrangement of the water molecules.

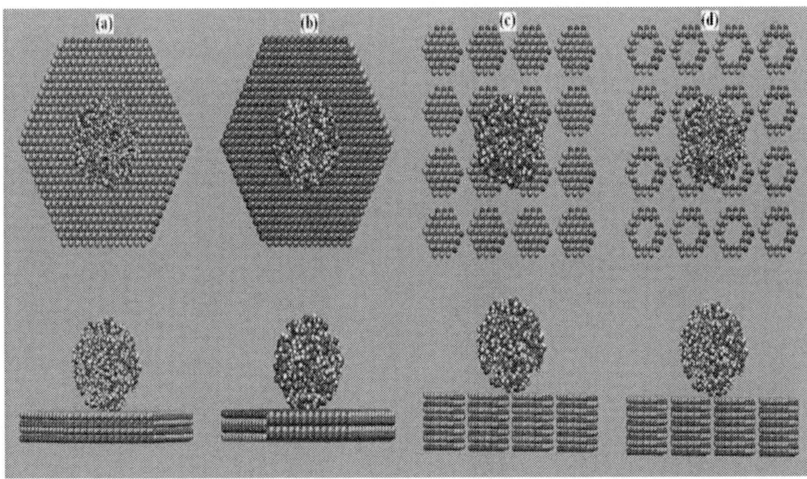

Figure 7. Top view and side view of the initial (t=0) water droplet (sphere) on Zn polar slab (a), O polar slab (b), ZnO nanorods array (c), and ZnO nanotubes array (d).

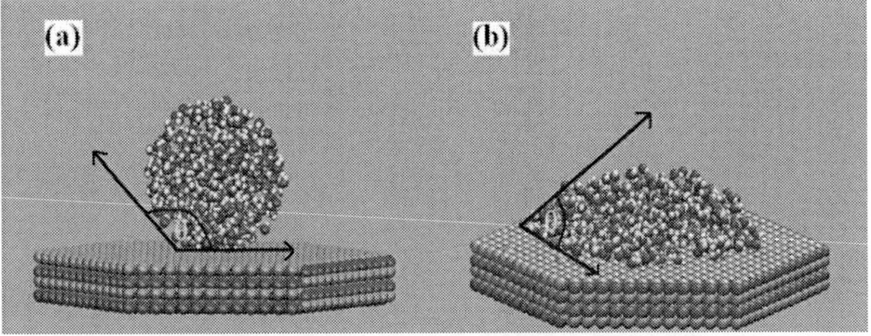

Figure 8. Side view of the initial (t=0) (a) and equilibrated (t=0.4 ns) (b) water droplet (sphere) on (0001)-Zn polar hexagonal slab.

The resulting voltage bias V across the simulated system depends on both the magnitude of the applied field E and the dimension L_z of the system in the direction of the field, i.e. $V = E L_z$ [30]. A constant electric field was applied to produce the desired drifting voltage across the system. The bias voltage was applied along (4.2 nm for the case of ZnO slab-water sphere, 4.0 nm for the case of ZnO slab-water box, 5.2 nm for the case of ZnO rod or tube-water sphere, and 5.0 nm for the case of ZnO rod or tube-water box) system length in z-direction (solution - ZnO nano structures) (see Figure 9). For the case of ZnO nanorods/tubes array we also approximate the problem to four nanorods/tubes system array to simulate the electrowetting to reduce the time of simulation, a water box with 40-38-30 Å^3 composed of 1412 water molecules was placed on top of four nanorods or nanotubes array (see Figure 10).

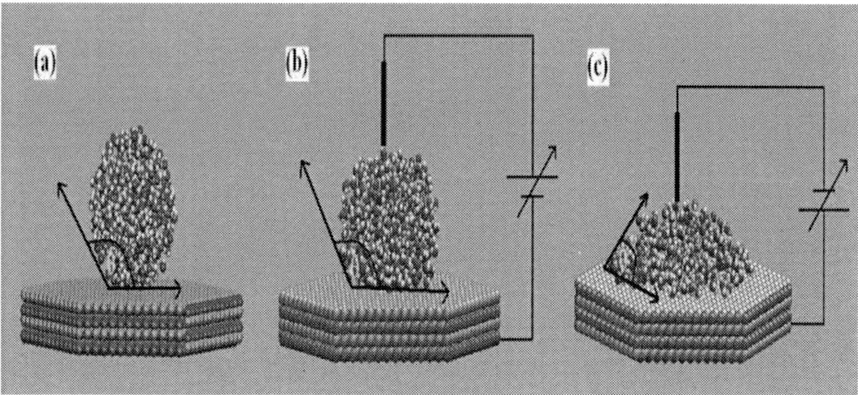

Figure 9. Side view of the initial (t=0) (a) and electro-wetted (t=0.6 ns) water droplet sphere on the (0001)-Zn polar slab at 20 V in the forward biasing (b) and -20 V in the backward biasing (c).

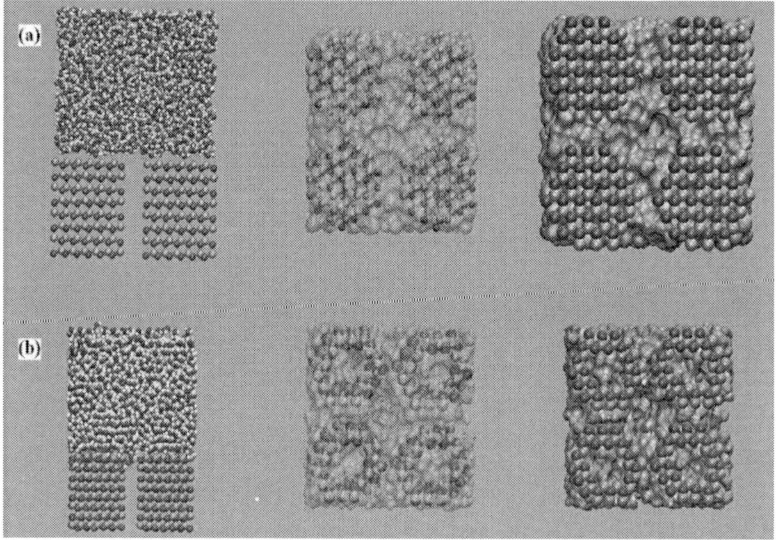

Figure 10. Initial array wetting systems with different viewing sides, consist of four ZnO nanorods array (a) or four ZnO nanotubes array (b).

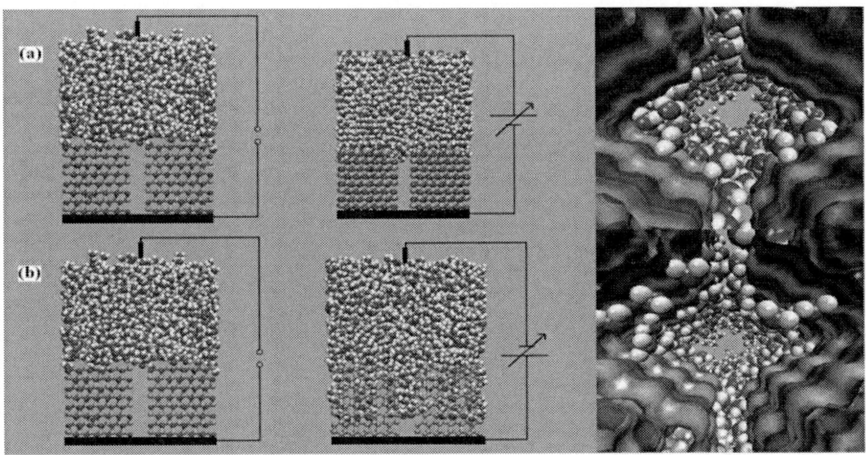

Figure 11. 20 V forward biasing for electrowetting of four ZnO nanorods array with different views (a) and 20 V backward biasing for four ZnO nanotubes array with different views (b), respectively. Nanorods case shows non-wetting behavior where in case of nanotubes it shows wetting behavior.

All systems were minimized for 1ps with time step 1 fs using the conjugate gradient method and then equilibrated for 2 ns in the *NVT* ensemble.

To keep the ZnO nanorod/tube rigid, the Zn and O atoms were either fixed or restrained to their original position by applying a harmonic force with a force constant of 10 kcal/mol/Å^2. Figure 11 illustrates the before and after the applied voltage. Samples of the trajectory are stored every 0.1 ps.

3.3.3. Water Permeation

We simulated the permeation of water through the ZnO nanotubes using the intermolecular parameters validated by our simulations. The ZnO nanotube was covered with a water box (20-20-30 Å^3) on each side then we cut the two boxes to hexagonal shape to allow us to use hexagonal periodic boundary conditions, and the resulting systems (having l_z=9.1nm) were equilibrated for 1 ns at 300 K under different conditions.

The number of water molecules within the nanotube was used as a measure of permeation (see Figure 12). A large number of simulations were run for water in model tubes order to investigate the influence of inner tube radius (0.12 nm $\leq$ R $\leq$ 1 nm) and inner tube surface character (hydrophobic vs. hydrophilic). To count the number of water molecules in the ZnO nanotube, we considered a cylindrical bin of 32 Å diameters concentric with the tube axis, with a height of 32 Å parallel to the Z-axis.

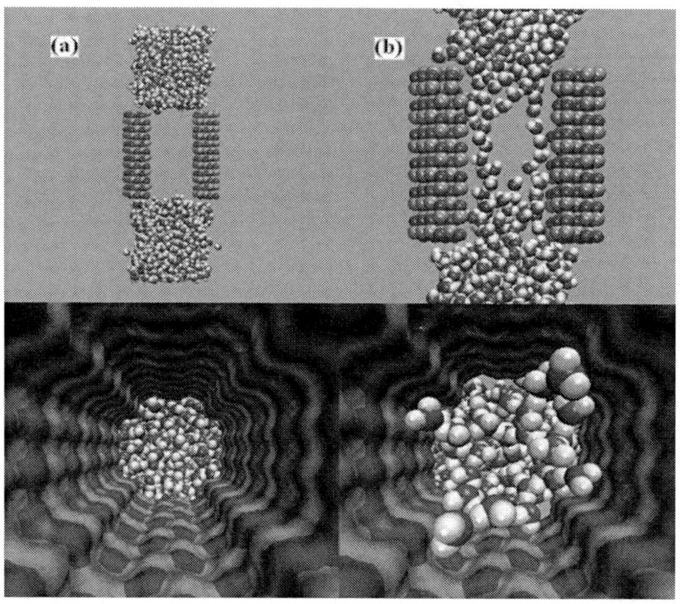

Figure 12. Initial at t=0 (a) and equilibrated at t=0.4 ns (b) water permeation through ZnO nanotube at T=300 K.

3.3.4. Ionic Currents

The simulation was performed for ZnO nanotube in presence of water molecules in contact with both top and base sides of the nanotube. The water is extended ~3 nm above and below the ZnO nanotube. Ions including Na^+, K^+, Ca^{2+}, Mg^{2+} and Cl^- were added to obtain a 10.0, 5.0, 1.0, 0.5, 0.1M sodium chloride (NaCl) or potassium chloride (KCl) or calcium chloride ($CaCl_2$) or magnesium chloride ($MgCl_2$) solutions (see Figure 13).

We have used the molecular dynamic simulations to perform ions transition through the ZnO nanotubes. To obtain a relative permittivity of 8.7 [31], harmonic restraints were applied to bulk ZnO atoms and harmonic bonds were applied between neighboring atoms. Similar forces were applied to surface atoms. In all simulations, hexagonal prism periodic boundary conditions were applied and non bonded energies were calculated using particle mesh Ewald full electrostatics [32] (grid spacing < 0.14 nm) and a smooth (1.0–1.2 nm) cutoff of the van der Waals energy.

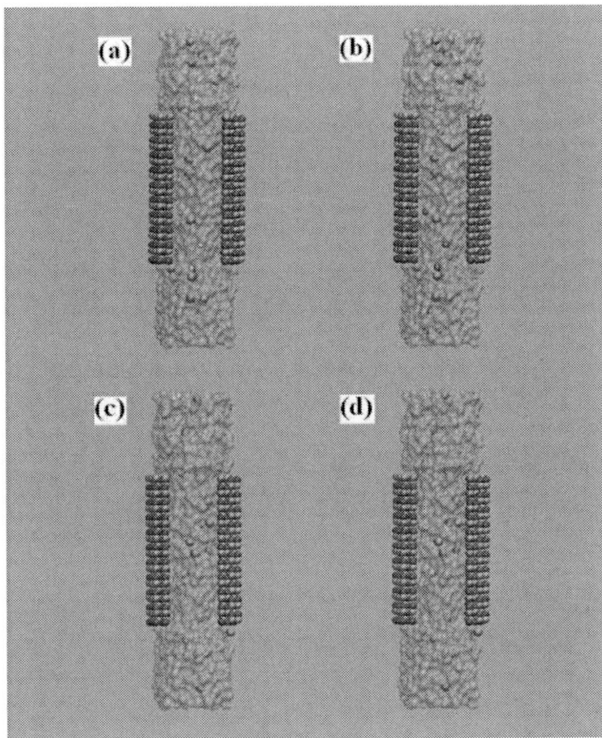

Figure 13. Filled ZnO nanotube with 1M concentration of (a) NaCl, (b) KCl, (c) CaCl$_2$, and (d) MgCl$_2$ electrolyte solutions.

Each system underwent 2 ps steps of energy minimization, 2 ps of equilibration at fixed volume. During this process the temperature increased from 0 to 300 K by rescaling of velocities, and the temperature was kept at 300 K by applying Langevin forces [33] to all atoms of the ZnO. The equilibration was performed using Nose'-Hoover Langevin piston pressure control at 1.0 atm for 0.5ns [34] with integration time step chosen was 1 fs.

All simulations, for both ZnO nanotube sizes, were performed at a fixed volume with the temperature maintained by Langevin dynamics applied only to the atoms of the ZnO. When a uniform electrical field is applied to all atoms, it induces, at the beginning of the simulation, a rearrangement of the ions and water focuses the electrical field to the vicinity of the ZnO nanotube. This has led to neutralizing the field in the bulk. The gradient of this electrostatic potential drives the ions current through the ZnO nanotube. A constant electric field was applied to produce the desired drifting voltage across the system. The bias voltage was applied along 11.1 nm system length in z-direction (solution - ZnO tube – solution). The present MD simulation allows a real time observation of the conformation and to determine the corresponding ionic current. The ionic current can be computed from the MD trajectory by summing up local displacements of all ions over a time interval between trajectory time frames interval Δt [35] as:

$$I\left(t + \frac{\Delta t}{2}\right) = \frac{1}{L_z \Delta t} \sum_{i=1}^{N} q_i \left(z_i(t + \Delta t) - z_i(t)\right)$$

(39)

where z_i and q_i are the z coordinate and charge of ion i, respectively, N is the total number of ions, and L_z is the system simulated length along the z axis in the direction of the applied field. The interval $z_i(t + \Delta t) - z_i(t)$ was computed respecting the periodic boundary conditions.

3.4. Water Density Profiles

From the MD simulation trajectories, water isochore profiles are obtained by introducing a cylindrical binning, which uses the topmost ZnO layer as zero reference level and the surface normal through the center of mass of the droplet as reference axis. The density profile was computed using horizontal layers of 1 Å heights. Each layer (bin) was displaced by 0.5 Å in the Z-direction (with equal volume), i.e., the radial bin boundaries are located at $r_i = \sqrt{i\, \partial A / \pi}$ for i=1, ..., Nbin with a base area per bin of $\partial A = 95$ Å2, starting from the base and ending at the top of the droplet. Within each layer, the radial density was calculated using circular bins of 1 Å thickness that were concentric around the center of mass of the droplet; for each bin, the inner boundary was displaced by 0.5 Å along the horizontal axis until 20 bins with zero density were reached. To extract the water contact angle from such a profile, a two-step procedure is adopted, as described by de Ruijter et al. [36]. First, the location of the equimolar dividing surface is determined within every single horizontal layer of the binned drop. Second, a circular best fit through these points is extrapolated to the ZnO surface where the contact angle θ is measured. Note that the points of the equimolar surface below a height of 5 Å from the ZnO surface are not taken into account for the fit, to avoid the influence from density fluctuations at the liquid-solid interface. Furthermore, only those points are used for which the density measured in the central bin lies within a range of 0.5-1.1 g cm^{-3}. This effectively excludes the points in the cap region where statistics are poor. The density profile from the bulk water region across the liquid-vapor interface ($z > 5$ Å) fits the hyperbolic tangent functional form which represents the decrease of the radial density:

$$\alpha(z) = \frac{\alpha_1}{2}\left(1 - \tanh\left(\frac{2(z - z_e)}{w}\right)\right)$$

(40)

where the vapor density is assumed to be zero, α_1 is the density of the bulk liquid, z_e the height (layer boundary), and w a measure for the width of the liquid-vapor interface (thickness of the interface). The layer boundary z_e resulting from matching Eq. 40 to the radial density was fitted to a circular segment, the boundary within 5 Å from the bottom being excluded.

4. RESULTS AND DISCUSSION

4.1. Density Profiles and WCA

After the ZnO polar surfaces were obtained and the size of the ZnO-water system was specified, we proceeded to investigate how the partial charge and force field parameters affect the wettability of different ZnO surfaces. Here, three series of MD simulations were considered within each of which the ZnO-water interaction parameters are identical, for different ZnO geometries. In the first series, (0001) and (000$\bar{1}$) ZnO polar surface slabs were used. For the second series, ZnO nanorods array were used. Finally, in the third series, ZnO nanotubes array were used.

In all the wetting and electrowetting simulations we used interaction potential with $D_{ZnO} = 0.3582$ $kJ\,mol^{-1}$ and $r_{ZnO} = 2.149$ Å and similar droplets size with a uniform density of 0.997 g cm-3. According to the modified Young's Eq. 3, the microscopic contact angle θ_w deviates from the macroscopic angle θ_Y due to the line tension τ. The effect of a positive line tension is to contract the droplet base and to increase the contact angle whereas a negative τ enhances wetting. However, as τ is believed to be on the order of 10-10 J/m, the line tension is only expected to be significant for droplets with diameters below 10 nm. The requirements for a precise experimental determination of τ include an accurate contact angle measuring technique and the use of a highly purified liquid on an atomically smooth surface. These requirements are perfectly matched in MD simulations. A straightforward method to determine the effect of the line tension τ in MD simulations is to measure the contact angle θ_w and the contact line curvature $1/r_B$ for droplets of different sizes. Figure 14 shows the time dependence during a 2 ns MD simulation of the WCA for a spherical droplet of 1428 water atoms resting on ZnO hexagonal polar slab surface. Similar time dependence was seen for all ZnO polar surfaces, showing an initial decrease within the first 1 ns, followed by an equilibrium state over the next 1 ns.

The WCA determined for the polar surfaces of ZnO slabs, it is seen to depend on the magnitude of the vdW interaction and is sensitive to the partial charges of zinc and oxygen atoms. In this case, the WCA indeed is seen to depend to the surface charge polarity magnitude where in case of O-polar surface the WCA is greater than 90° (hydrophobic) and in case of Zn-polar surface is less than 90° (hydrophilic) [37].

The results described are consistent with explanations of partial charge at the surface, like dangling Zn atoms and dangling O atoms, act as hydrophilic and hydrophobic centers, respectively.

Fully coordinated atoms are partially buried at the surface; hence, their electrostatic contribution is effectively screened by their neighbors.

Conversely, dangling atoms are more exposed and the charges of q_O and q_{Zn} produce oriented dipoles that act as adsorption sites for water. The calculation of the water contact angle is schematically presented in Figure 15.

The density profile was computed using fitting plot (were done using the fit command and interpolation provided by MatLab v. 7.8). The WCA was averaged over the last 0.2 ns for each MD simulation (see Figure 14). A few water molecules eventually evaporated from the droplet; water molecules 5 Å away from the droplet were not taken into account. It was predicted that these parameters lead to a microscopic contact angle of 20° for a 1428 water atoms droplet (TIP3P water) on (0001)-ZnO polar slab. The simulation results in a contact angle of 22.6°, which is in good agreement with the expected value [38,39]. Note that the layered structure of the liquid close to the wall ($0 < z < 5$ Å) is neglected in the contact angle measurement. A density profile along the centerline of the droplet is shown in Figure 16.

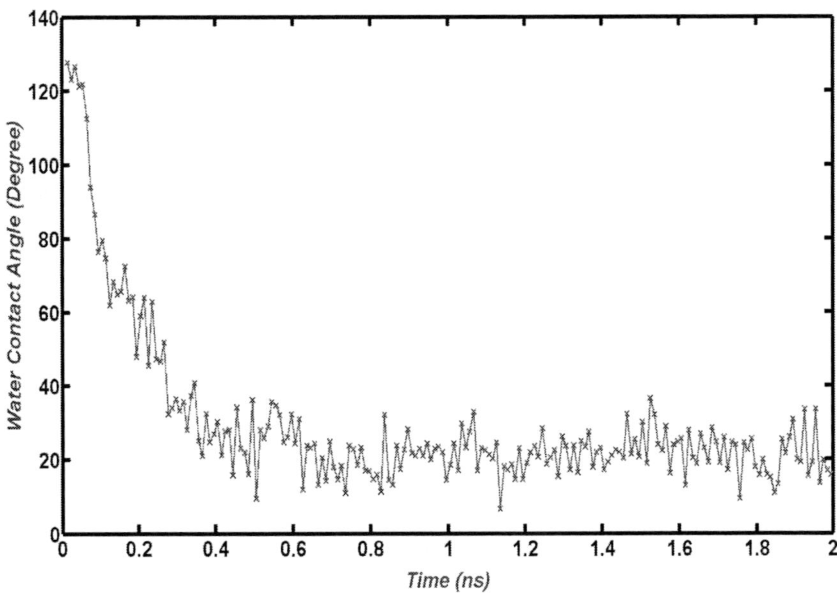

Figure 14. Variation of the contact angle during an equilibration of a water droplet on (0001)-ZnO polar slab where the initial water droplet shape is sphere with diameter of 32 Å placed on the top. The WCA values during the last 0.2ns were averaged and used as an equilibrium value.

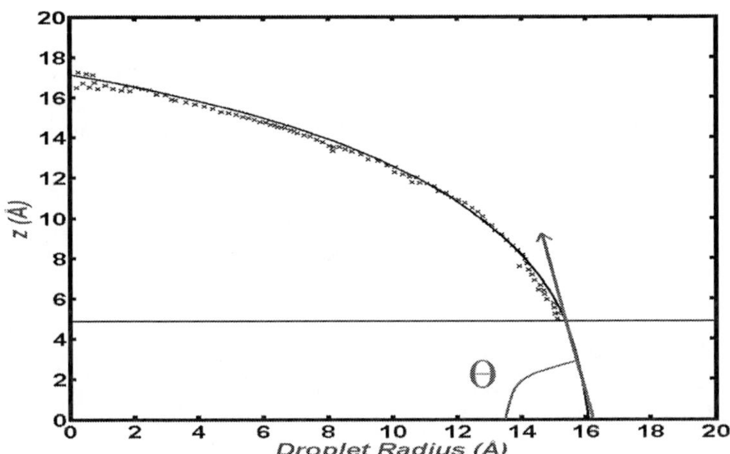

Figure 15. Water contact angle on (0001)-ZnO polar slab measured by fitting a circle with center (0,z) and radius r to the points of the equimolar planes with z>zo=5 Å from the ZnO surface is excluded from the calculations.

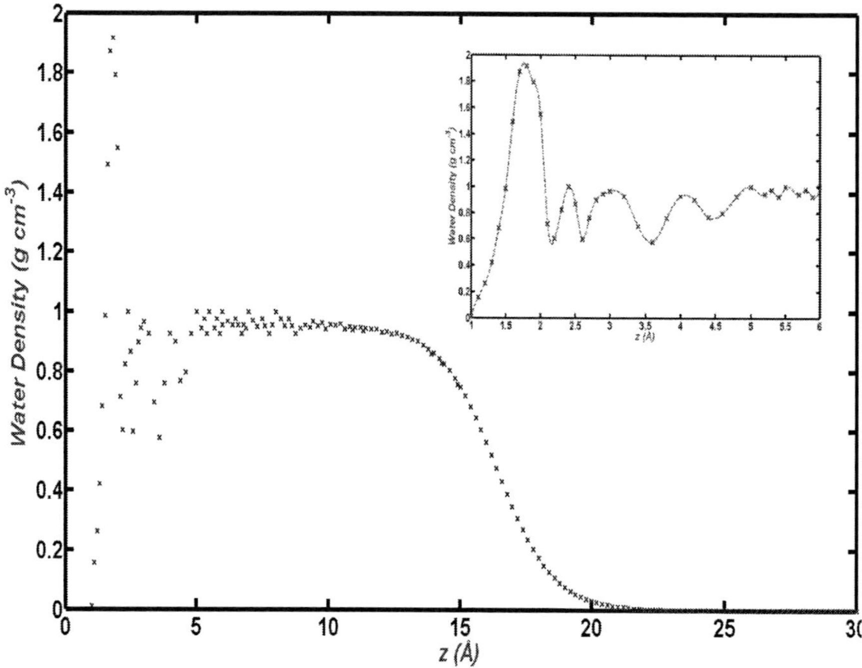

Figure 16. Water density profile for (0001)-ZnO polar slab along the centerline of the droplet; for the calculation of the contact angle only the data points contained within the boundaries ($z>5$ Å : $0.5<\alpha_1(z)<1.1$ g cm^{-3}).

Close to the ZnO, one pronounced density peak can be identified at distance of 1.8 Å with peak height of 1.9 $g.cm^{-3}$. A fit of Eq. 40 to the profile obtained for (0001)-ZnO polar slab yields a bulk liquid density of $\alpha_1 = 0.989$ $g.cm^{-3}$ and $w = 4.5$ Å, resulting in an interface thickness of 4.9 Å (the interface thickness is here defined as the region where the water density drops from 0.9 α_1 to 0.1 α_1).

For the electrowetting case, we have varied the applied voltage on the system to find the values that reproduce the WCA. In this case, the positive partial charge of the Zn surface atoms on (0001)-ZnO polar slab were modulated according to the polarity of the applied voltage (+20 V) making the (0001)-Zn surface react as hydrophobic surface with WCA of 96.3° (see Figure 17) [40].

Employing these applied potentials to the electrowetting model problem of a water droplet on ZnO polar slab results in qualitatively different behavior ranging from a strongly hydrophobic to hydrophilic interface [40,41] (see Figure 9).

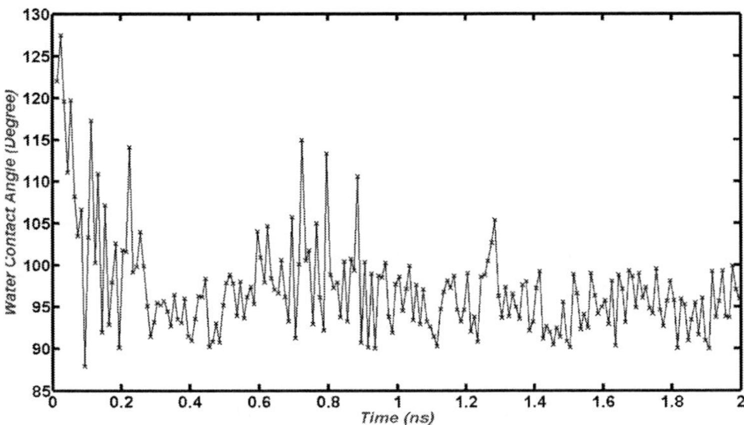

Figure 17. Variation of the contact angle during application of 20 volts on the system (water droplet on (0001)-ZnO polar slab) where the initial water droplet shape is sphere with diameter of 32 Å placed on the top. The WCA values during the last 0.2ns were averaged and used as an equilibrium value.

These ZnO surface behaviors for switching for hydrophobic to hydrophilic faces were also are found in case of simulating the wetting and electrowetting behaviors for ZnO nanorods/tubes array with water box droplet, here the wetting mechanisms have two behaviors in the same time, spreading which is small compared with flat surfaces and filling which is depending on the nanorod size and the spaces between the nanorods/tubes for the organized array (result not shown) [42-48,40].

4.2. Water Permeation through ZnO Nanotube

The water density in nanotube oscillates between liquid and vapour on a nanosecond time scale, a manifestation of capillary evaporation and condensation at the nanoscale. For hydrophobic tube a strong dependence of the tube state on the radius is apparent.

The stable thermodynamic state switches from vapor to liquid at a critical radius $r_c = -L\cos\theta_Y$ (using $\Delta U(r_c) = 0$ in Eq. 22). The coefficient of the quadratic term, $\gamma_{lv} + (\tfrac{1}{2})\,\Delta\mu\,\Delta b_{vl}\,L$, is positive for the hydrophobic tubes, consistent with the model Eq. 21, which predicts this coefficient to be independent of the tube inner wall. For the polar tube show that a high density of local charges leads to a higher probability of the tube being liquid filled than predicted by the macroscopic model.

For the system parameters, the coefficient is in fact dominated by the water liquid vapor surface tension γ_{lv}. $\Delta\gamma_s$, the difference in surface tensions between the tube inner wall and vapor or liquid, becomes more positive with increasing polarity of the tube wall. It effectively measures the hydrophobicity of the tube inner wall. This becomes even more apparent when the (macroscopic) contact angle $\cos\theta_Y = \Delta\gamma_s / \gamma_{lv}$ is formally computed. Macroscopically, a hydrophobic surface can be defined as one with $\theta_Y > 90^o$. A "hydrophilic" tube system is characterized by $\Delta\gamma_s > 0$ or $\theta_Y < 90^o$ and liquid is always the preferred phase in the inner tube, regardless of inner tube radius [18].

Further tests of Eq. 22 by varying the length of the tube together with the radius confirm the model qualitatively (data not shown). Our model implies that for nanoscale tubes the cost of creating the liquid-vapor interface is the only force driving the filling of a hydrophobic ($\Delta\gamma_s \leq 0$) tube. Little free energy $\Delta\mu\,\Delta b_{vl}\,\pi\,r^2 L$ is gained by creating a bulk-like liquid in the tube instead of vapor.

In the electro-simulations we control the applied voltage; the system was simulated for 1 ns. In the forward biasing first case, water completely permeated the nanotube; in the backward biasing second case, water did not penetrate the tube completely. We conclude that ZnO nanotube wall act as a capacitor inside the nanotube which can be responsible for the initial water permeability of the tube; rather the surface hydrophobicity slows down the wetting. To further test the effect of the surface hydrophobicity on wetting, we increased the hydrophobicity of the ZnO nanotube by increasing the applied voltage and analyzed the permeation kinetics. In this case, we observe that Zn and O surface atoms affect the permeation speed. Furthermore, the velocity of the permeation is faster for tubes with a larger diameter. In these two cases, the permeation behavior can be explained in terms of cohesive (water-water) and adhesive (water-ZnO) forces [19].

4.3. Salt Concentration Dependence on ZnO Nanotube Ionic Currents

In principle, three mechanisms contribute to electrolyte transport through the nanotube, these are: electrophoretic ion migration, convection (i.e., electro-osmosis), and diffusion. However, the contributions of convection and diffusion are marginal, and the contribution from ion migration predominates.

As a first step toward understanding the modulation of ionic conductivity, we simulate ion transport in hexagonal ZnO nanotube with cylindrical inner shape having radius 1.46 *nm* by using MD simulations. We further calculate the ionic current as a function of the applied biasing voltage for constant electrolyte concentrations and the ionic conductance as a function of the electrolyte concentration for a constant tube inner diameter. The intention is to infer the conductivity of the electrolyte solution for the case of electrolyte solution flow in ZnO nanotube with fixed surface charges.

The dc electrolytic current through a single ZnO nanotube is calculated as a function of the applied electrochemical potentials for 20, 15, 10, 5, 4, 3, 2, 1, -1, -2, -3, -4, -5, -10, -15, -20 volt in the z-direction at 27 °C. Starting from a random configuration, the systems were simulated for 0.5 ns to reach a steady state, followed by a run of 2 ns.

The ion current was computed using Eq. 39. The current–voltage (I–V) characteristics obtained from ZnO tube with inner radius of 1.46 nm, calculated over a range of electrolyte concentrations 0.5, 1.0, 5.0, 10.0 M. Notice that in all cases, the I–V characteristics are approximately linear which means that they have ohmic behavior (see Figure 18a, b). By fitting a line to the data in Figure 18 and taking the slope we can obtain the conductance (see Figure 19), which is function of ion concentrations for a constant ZnO nanotube diameter.

However, it is apparent that the conductance increased exponentially with increasing the ion concentration to 5M, after this concentration value the conductance appear to have stable behavior. It is appear from the Figure 19 that the highest ionic conductance is for KCl solution then NaCl, $CaCl_2$ and $MgCl_2$ which are in a good agreement with the ionic size for these ions [49].

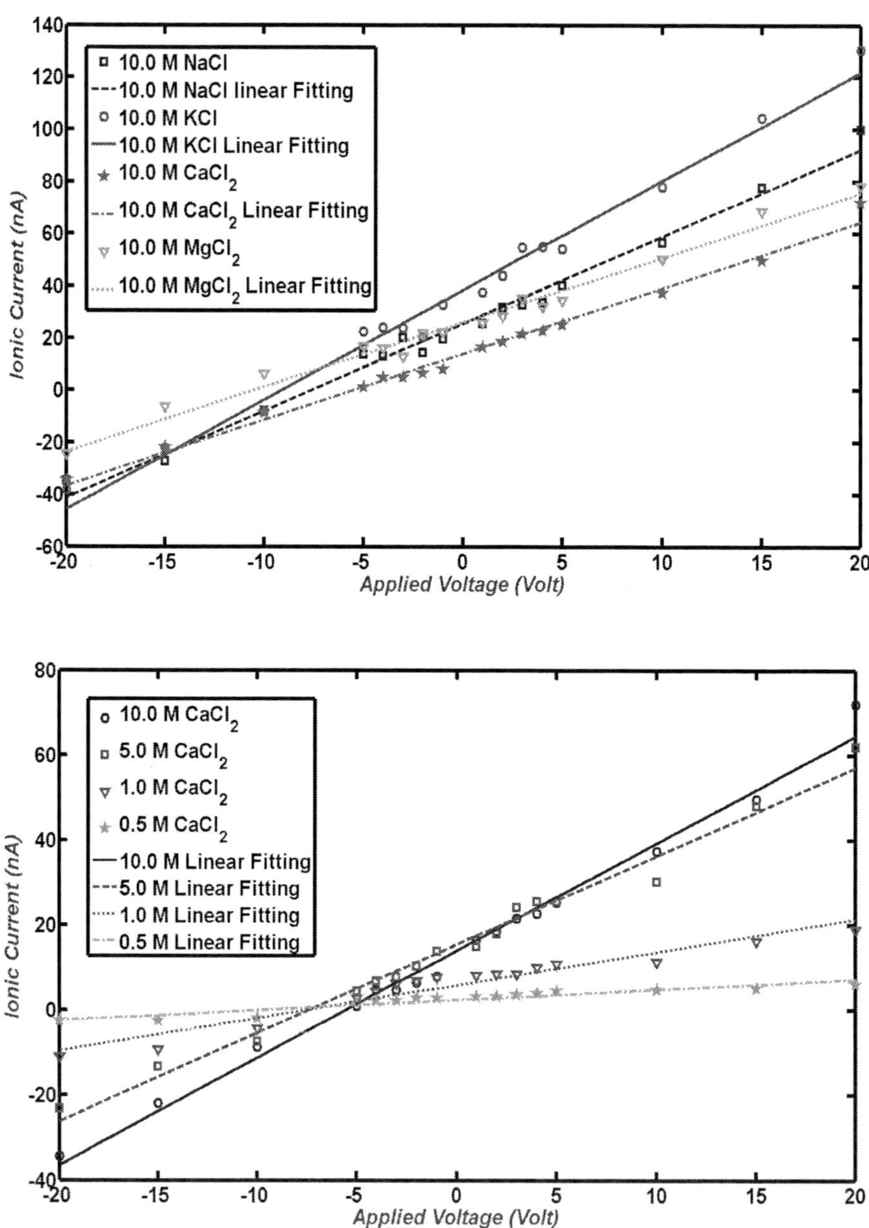

Figure 18. Ionic current as a function of applied bias voltage for (a) 10M electrolyte concentration of NaCl, KCl, $CaCl_2$, and $MgCl_2$ and for 0.5, 1.0, 5.0, and 10.0 M $CaCl_2$ (b).

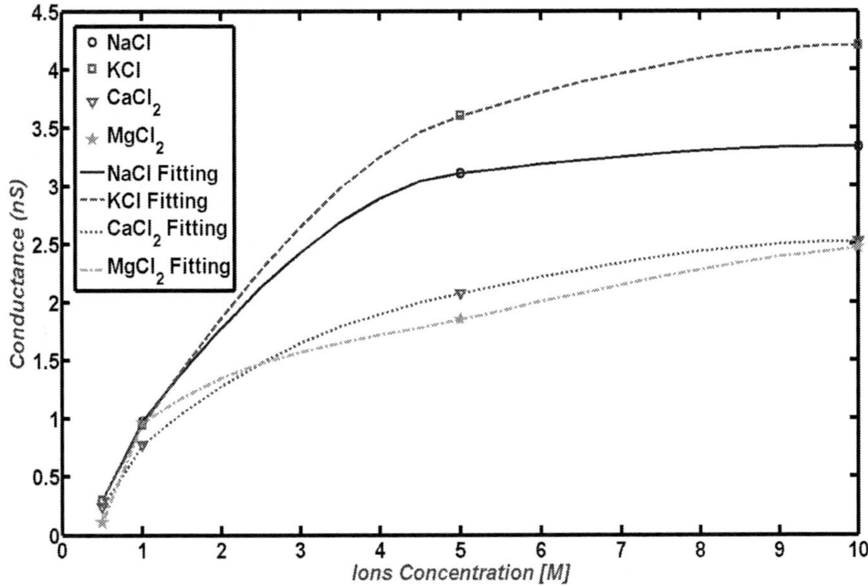

Figure 19. Ionic conductance as a function of ion concentration for NaCl, KCl, CaCl$_2$, and MgCl$_2$ electrolyte solutions.

In the MD simulations we observed adhesion of flowing ions to the top, the bottom and to the inner surfaces of the ZnO nanotube. This adhesion will lead to reduce the flowing ionic current value. The observed adhesion was dominated by hydrophobic attraction of the nanotube walls with the water molecules and ions creating the interface region and the double layer capacitance. Figure 20 shows snap shot images of the ions movement in the ZnO nanotube (r_{tube}=1.65 nm) obtained from the MD simulations, indicating that the positions and the amount of ions for 2M KCl are adsorbed on the (000 $\bar{1}$)-O and (0001)-Zn permanent polar surfaces, respectively.

Both of them were adsorbed on the dangling bonds of the nanotube inner wall surfaces depending on the electrolyte concentrations and the applied voltage. It is important to note that in our case we have a nanotube with a very thin thickness so that the permanent polar surface charge value can be reduced to include approximately the top and bottom edges of the inner tube surface.

It is known that ZnO nanotubes surfaces carry a small net charge [50,51], the partial charge on the surface atoms can generate localized strong electric field that attracts ions.

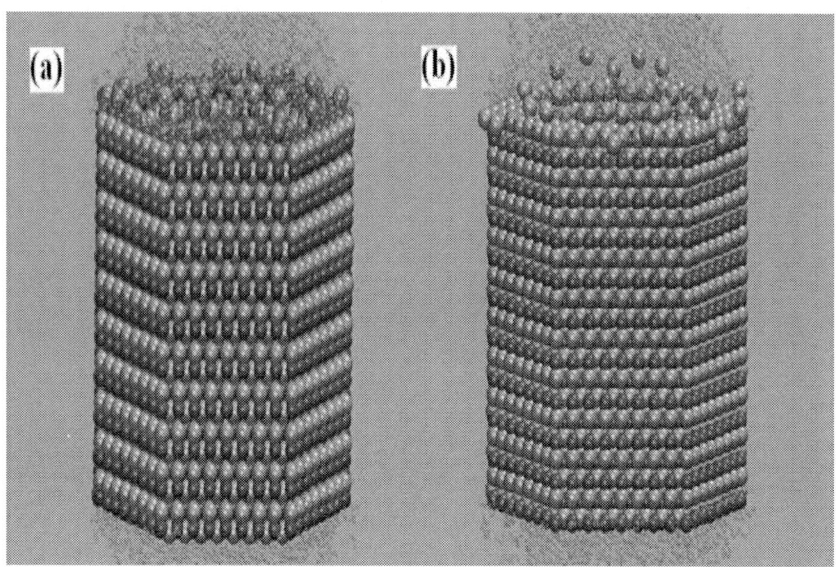

Figure 20. Adsorptions of ions on the O-polar surface (a) and Zn-polar surface (b) during the MD simulations run for 2M KCl electrolyte solution [53].

To understand the scaling behavior of the ionic conductance, changes in the electrolyte concentration and ionic mobility have to be taken into account. To calculate the mobility of cation μ_{c^+} and anion μ_{a^-} we used the following equations [21, 52]:

$$N_{c^+} = \frac{1}{e\mu_{c^+}\rho} \tag{41}$$

$$N_{a^+} = \frac{1}{e\mu_{a^+}\rho} \tag{42}$$

where N_{c^+} and N_{a^-} are the number of cation and anion, and ρ is the resistivity of electrolyte solution channel under zero gate voltage is determined from [21]:

$$\rho = \frac{dV_{DS}}{dI_{DS}}\frac{A}{L} \tag{43}$$

where A is the channel cross-section area and L is the length of the channel between the source and drain electrodes. The adsorption of cation and anion at the inner wall surface over screens the surface charge. Notice that the cation and anion concentration are oscillates with distance from the inner wall of the tube.

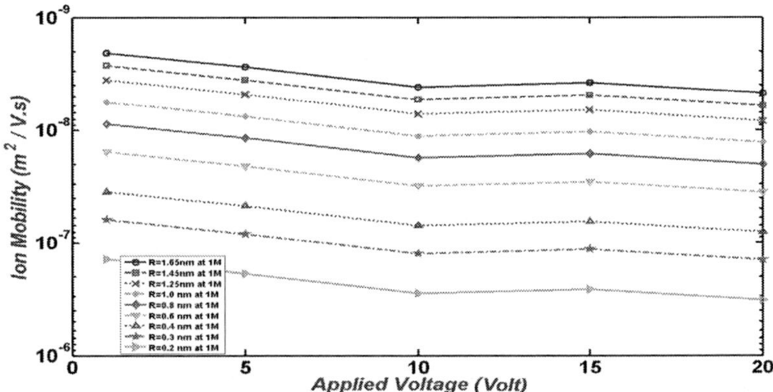

Figure 21. Ion mobility as a function of the applied bias voltage for ZnO nanotube R_{in}=1.65nm at 1M KCl with changing the distance from the tube inner surface [53].

The ion mobility gradually increases as it moves away from the inner tube surface toward the center of the tube. In the larger tube inner diameter (D_{in} = 3.3 nm), the mobilities near the center of the tube are almost the same and are comparable with the bulk value, whereas for smaller inner diameter (D_{in} = 2.0 nm) tube, the ions mobility in the tube center is smaller of its bulk value (see Figure 21). The reduction in the ion mobilities are observed, indicating that the ion mobility in a nanotube with inner diameter D_{in} = 2.0 nm is substantially affected by confinement and ion–surface interactions [53].

CONCLUSION

We have explicitly demonstrated a systematic study of the potential functions used to model the interaction between water and ZnO nanostructures in molecular dynamics simulations.

Our results were performed to analyze the behavior of water and ions adsorbed on ZnO surfaces on the nanosecond scale. Three different geometries were considered in the study: (0001)-Zn and (000$\bar{1}$)Zn-O polar surface slabs, nanorod crystals, and nanotubes were modeled. We have used UFF force field parameters and refined them to fit the CHARMM force field parameters for ZnO surfaces that simulate wetting properties with different ZnO nanostructure films based on the observation of WCA. The water density profiles were analyzed and yielded important results. A hydrophobic mechanism was found for water droplet interaction with (000$\bar{1}$)-ZnO polar slab with contact angle 96° (water moves away from the surface) and a hydrophilic mechanism for the (0001)-ZnO polar slab with contact angle 54° (overlap between water and the surface atoms) for the equilibrium case as results of partial surface charges interactions acting between the oxygen atoms of the water and Zn atoms sites for (0001) surface or O atoms sites for (000$\bar{1}$) surface. An applied voltage on the system were used to simulate the electrowetting behavior, these interactions shows, for the case of (0001)-ZnO polar slab when we applied backward biasing voltage, the system still have hydrophilic face with WCA less than 20°, on the other hand when we forward biasing voltage the system will have hydrophobic face with WCA of 96.3°. These ZnO surface behaviors for switching from hydrophobic to hydrophilic faces were also founded in cases of simulated the wetting and electrowetting behaviors for ZnO nanorods/tubes array where the wetting mechanisms have two behaviors in the same time, spreading and filling.

The results from the permeation of water through ZnO nanotube model are an important factor to explain the initial hydrophobicity and hydrophilicity of the ZnO nanotubes for equilibrium and applied voltage cases. A thermodynamic model based on surface energies fits remarkably well with the data of the atomic-scale MD simulations. Our simulations show that the polarity of the tube wall can shift the WCA considerably. By applying voltage across the system the hydrophobic face are rotated to hydrophilic face by combining hydrophobic face with a change in surface polarity. It appears that in this case water structure (i.e. a hydrogen bond network) is responsible for different wall-fluid surface tensions. As a result, a moderate change in applied voltage is required to obtain the required face behavior for different applications.

We have inferred the conductance and ionic mobility of NaCl, KCl, $CaCl_2$, and $MgCl_2$ electrolytes flowing through ZnO nanotubes. We performed calculations of the ionic current as a function of the electrolytes concentration. The nanotube conductance variation versus the concentrations are found to change nonlinearly with increasing the electrolyte concentration and after 5M the conductance have reached saturation values for the nanotube diameter (3.2 nm).

We interpret these observations by using MD simulations of the ion transport through the ZnO nanotubes and found that the calculated conductance are consistent with the presence of fixed surface charges in the nanotube inner surface. Moreover, a reduction of the ion mobility due to this surface charge is also observed. We conclude that the ionic current in ZnO nanotube are governed by fixed surface charge layer and can be modulated by applying voltage across the ZnO nanotube outer walls (V_{gate}).

In this case a ZnO nanotube can be operated in a way analogous to FET operation leading to control of ionic fluid flow through nanotubes [54-56]. Such control of ionic flow would be of interest to many biological systems. This model will be used to simulate the ion selective conductance through ZnO nanotubes and to calculate the binding infinities of various biological macromolecules to ZnO surfaces.

REFERENCES

[1] Kumar, S. A.; Chen, S.-M. *Analyt. Lett.* 2008, *41*, 141-158.

[2] Wei, A.; Sun, X.W.; Wang, J.X. *Appl. Phys. Lett.* 2006, *89*, 123902.

[3] Yeh, P.-H.; Li, Z.; Wang, Z. L. *Adv. Mater.* 2009, *21*, 1–4.

[4] Yang, K.; She, G.-W.; Wang, H.; Ou, X.-M.; Zhang, X.-H.; Lee, C.-S.; Lee, S.-T. *J. Phys. Chem. C*, 2009, DOI: 10.1021/ jp901894j.

[5] Al-Hilli, S.; Willander, M. *Nanotechnology* 2009, *20*, 175103.

[6] Al-Hilli, S. M.; Öst, A.; Strålfors, P.; Willander, M. *J. Appl. Phys.* 2007, *102*, 084304.

[7] Al-Hilli, S.; Willander, M. *Sensors* 2009, *9*, 7445-7480.

[8] Young, T. *Philos. Trans. R. Soc. London* 1805, *95*, 65-87.

[9] Wang, J. Y.; Betelu, S.; Law, B.M. *Phys. Rev. E* 2001, *63*, 031601.

[10] Adamson, A. W. *Physical Chemistry of Surfaces*; 6th edit; John Wiley and Sons: New York, 1997, pp 195-197.

[11] Welters, W. J. J.; Fokkink, L. G. J. *Langmuir* 1998, *14*, 1535-1538.

[12] Mugele, F.; Baret, J.-C. *J. Phys.: Condens. Matter* 2005, *17*, R705–R774.

[13] Wenzel, R. N. *Ind. Eng. Chem.* 1936, *28*, 988–994.

[14] Cassie, A. B. D.; Baxter, S. *Trans. Faraday Soc.* 1944, *40*, 546 – 551.

[15] Han, J.; Gao, W. *J. Electron. Mater.* 2009, *38*, 601-608.

[16] Beckstein, O.; Sansom M. S. P. *Phys. Biol.* 2004, *1*, 42-52.

[17] de Gennes, P. G. *Rev. Mod. Phys.* 1985, *57*, 827–863.

[18] Allen, R.; Hansen, J. P.; Melchionna S. *J. Chem. Phys.* 2003, *119*, 3905–3919.

[19] Heikenfeld, J.; Zhou, K.; Kreit, E.; Raj, B; Yang, S.; Sun, B.; Milarcik, A.; Clapp, L.; Schwartz, R. *Nature Photonics* 2009, *3*, 292-296.

[20] Vidal, J.; Gracheva, M. E.; Leburton, J.-P. *Nanoscale Res. Lett.* 2007, *2*, 61–68.

[21] Sze, S. M. *Physics of Semiconductor Devices*; Wiley: New York, 1981.

[22] Phillips, J. C.; Braun, R.; Wang, W.; Gumbart, J.; Tajkhorshid, E.; Villa, E.; Chipot, C.; Skeel, R. D.; Kale, L.; Schulten, K. *J. Comput. Chem.* 2005, *26*, 1781-1802 http://www.ks.uiuc. edu/ Research/namd/.

[23] Matlab, v. 7.5; *The MathWorks*, Inc., 2007.

[24] Humphrey, W.; Dalke, A.; Schulten, K. *J. Molec. Graphics* 1996, *14*, 33-38 http://www.ks.uiuc.edu/Research/vmd/.

[25] Rappé, A. K.; Casewit, C. J.; Colwell, K. S.; Goddard, W. A. I.; Skiff, W. M. 1992 *J. Am. Chem. Soc.* 1992, *114*, 10024-10035.

[26] MacKerell, A. D.; Brooks, B.; Brooks III, C. L.; Nilsson, L.; Roux, B.; Won, Y.; Karplus, M. *CHARMM: The energy function and its parameterization with an overview of the program*. In *The Encyclopedia of Computational Chemistry*; P. Schleyer; Ed.; John Wiley and Sons: UK, 1998, pp. 271–277.

[27] Schulz, H.; Thiemann, K. H. *Solid State Commun.*1979, *32*, 783.

[28] Werder, T.; Walther, J. H.; Jaffe, R. L.; Halicioglu, T.; Koumoutsakos, P. *J. Phys. Chem. B* 2003, *107*, 1345-1352.

[29] Cruz-Chu, E. R.; Aksimentiev, A.; Schulten, K. *J. Phys. Chem. B* 2006, *110*, 21497-21508.

[30] Martyna, G. J.; Tobias, D. J.; Klein, M. L. *J. Chem. Phys.* 1994, *101*, 4177–4189.

[31] Shimomura, K.; Nishiyama, K.; Kadono, R. *Phys. Rev. Lett.* 2002, *89*, 255505.

[32] Batcho, P. F.; Case, D. A.; Schlick, T. *J. Chem. Phys.* 2001, *115*, 4003–4018.

[33] Brunger, A. T. *X-PLOR*, Version 3.1: *A System for X-ray Crystallography and NMR*; The Howard Hughes Medical Institute and Department of Molecular Biophysics and Biochemistry, Yale University, 1992.

[34] Martyna, G. J.; Tobias, D. J.; Klein, M. L. *J. Chem. Phys.* 1994, *101*, 4177–4189.

[35] Aksimentiev, A.; Schulten, K. *Biophys. J.* 2005, *88*, 3745–3761.

[36] de Ruijter, M. J.; Blake, T. D.; De Coninck, J. *Langmuir* 1999, *15*, 7836-7847.

[37] Wu, P. C.; Losurdo, M.; Kim, T.-H.; Giangregorio, M.; Bruno, G.; Everitt, H. O.; Brown, A. S. *Langmuir* 2009, *25*, 924-930.

[38] Zhang, Z.; Chen, H.; Zhong, J.; Saraf, G.; Lu, Y. *J. Electron. Mater.* 2007, *36*, 895-899.

[39] Luo, J. *Langmuir* 2005, *21*, 7358-7365.

[40] Campbell, J. L.; Breedon, M.; Latham, K.; Kalantar-zadeh K. *Langmuir* 2008, *24*, 5091-5098.

[41] Shamai, R.; Andelman, D.; Bergec, B.; Hayes, R. *Soft Matter* 2008, *4*, 38–45.

[42] Li, G.; Chen, T.; Yan, B.; Ma, Y.; Zhang, Z.; Yu, T.; Shen, Z.; Chen, H.; Wu, T. *Appl. Phys. Lett.* 2008, *92*, 173104.

[43] Zhang, J.; Huang, W.; Han, Y. *Langmuir* 2006, *22*, 2946-2950.

[44] Sun, M.; Du, Y.; Hao, W.; Xu, H.; Yu, Y.; Wang, T. *J. Mater. Sci. Technol.* 2009, *25*, 53-57.

[45] Wu, X.; Zheng, L.; Wu, D. *Langmuir* 2005, *21*, 2665-2667.

[46] Zhang, X.-T.; Sato, O.; Fujishima, A. *Langmuir* 2004, *20*, 6065-6067.

[47] Feng, X.; Roy, S. C.; Grimes, C. A. *Langmuir* 2008, *24*, 3918-3921.

[48] Feng, X.; Feng, L.; Jin, M.; Zhai, J.; Jiang, L.; Zhu, D. *J. Am. Chem. Soc.* 2004, *126*, 62-63.

[49] Zumdahl, S. S. *Chemical Principles*; 5th Edit.; Houghton Mifflin Company: US, MA, 2005.

[50] Wander, A.; Harrison, N. M. *J. Chem. Phys.* 2001, *155*, 2312-2316.

[51] Kresse, G.; Dulub, O.; Diebold, U. *Phys. Rev. B* 2003, *68*, 245409.

[52] Muller, R. S.; Kamins, T. I.; Chan, M. *Device Electronics for Integrated Circuits*; Wiley: New York, 2003.

[53] Al-Hilli, S.; Willander, M. *Nanotechnology* 2009, *20*, 505504.

[54] Karnik, R.; Fan, R.; Yue, M.; Li, D.; Yang, P.; Majumdar, A. *Nano Lett.* 2005, *5*, 943-948.

[55] Vermesh, U.; Choi, J. W.; Vermesh, O.; Fan, R.; Nagarah, J.; Heath, J. R. *Nano Lett.* 2009, *9*, 1315-1319.

[56] Lee, C.-S.; Kim, S. K.; Kim, M. *Sensors* 2009, *9*, 7111-7131.

In: Molecular Dynamics
Editors: D. E. Garcia and P. J. Green

ISBN: 978-1-62081-545-8
© 2012 Nova Science Publishers, Inc.

Chapter 5

MOLECULAR DYNAMICS SIMULATIONS OF LIQUID AND IONIC SOLVATION OF CARBON TETRACHLORIDE: BASED ON ATOM-BOND ELECTRONEGATIVITY EQUALIZATION METHOD FUSED INTO MOLECULAR MECHANICS (ABEEM/MM)[*]

Xin Li, Li-Dong Gong and Zhong-Zhi Yang[†]
School of Chemistry and Chemical Engineering,
Liaoning Normal University, P. R. China

ABSTRACT

A flexible body and nine-site fluctuating charge model to describe the intramolecular and intermolecular interactions for carbon tetrachloride is constructed based on the combination of atom-bond electronegativity equalization method and molecular mechanics (ABEEM/MM). The potential parameters are refined to accurately

---

[*] A version of this chapter also appears in Advances in Chemical Modeling, edited by Mihai V. Putz, published by Nova Science Publishers, Inc. It was submitted for appropriate modifications in an effort to encourage wider dissemination of research.

[†] E-mail address: zzyang@lnnu.edu.cn. (To whom correspondence should be addressed)

describe the structures and binding energies of $(CCl_4)_2$ and $Na^+(CCl_4)_n$ (n=1-4) clusters.

We then carried out liquid CCl_4 and Na^+salvation in liquid CCl_4 simulations to examine the potential parameters. The computed density of liquid CCl_4 is in excellent agreement with experimental value.

The structures of liquid CCl_4 and Na^+ in liquid CCl_4 can be analyzed by examining the radial distribution functions and angular distribution functions.

It is found that the liquid CCl_4 forms an interlocking structure and that a local orientation correlation is observed between neighboring CCl_4 molecules. In the study of Na^+ solvation in liquid CCl_4, we observe a well-defined solvation shell around Na^+ with five coordinated CCl_4 molecules. It is also found that Na^+ induces a strong local orientation order in liquid CCl_4. Moreover, the calculated diffusion constant of CCl_4 is in reasonable agreement with the experimental result. This work demonstrates that the new potential model is good enough to reproduce properties of liquid $CCl_4$and Na^+ in bulk liquid CCl_4.

Keywords: ABEEM/MM; carbon tetrachloride liquid; ionic solvation; molecular dynamics simulation

1. INTRODUCTION

Molecular dynamics (MD) simulations have proven to be very useful tools for examining the condensed-phased properties [1-3]. The simulation studies are not only able to characterize the static structure of the investigated system at molecular level, but also to describe the dynamical processes such as solvation and ion transport occurring in the media. The accuracy of simulations is dependent on the potential model describing the interaction between molecules in the system. A careless choice of the interaction potentials can often lead to misleading conclusion.

Thus, it is necessary to construct the model potential that can properly describe the molecular interactions.

In this study, we develop a new potential model for carbon tetrachloride (CCl_4). CCl_4, being analogous to simple atomic liquids due to its high molecular symmetry, can serve as a simple model system for the under-standing of more complex molecular liquids, and it is also a very important organic solvent, widely used in the processes of synthesis, purification, and cleaning. Due to the usage of CCl_4 in the processes of treating radioactive materials, it contributes significantly to the nuclear waste remediation.

Therefore, much experimental and theoretical work has been devoted to the understanding of its chemical and physical properties [4-13], and it is important to develop a potential model that can describe the CCl_4 interactions properly, so one can study processes such as ion transport and chemical reactions in this solvent. There are several available CCl_4 interaction potentials in the literature [9-14], which are able to reproduce well the thermodynamic and structural properties of CCl_4. However, they are all fixed charge and rigid body potential models. In the present study, we will focus on constructing a nine-site fluctuating charge and flexible body CCl_4 potential, based on the atom-bond electronegativity equalization method fused into molecular mechanics (ABEEM/MM) [15-21].

The potential parameters are optimized by fitting the structural properties and binding energies of dimer $(CCl_4)_2$ and $Na^+(CCl_4)_n$ (n=1-4) clusters. This potential is then applied to model the liquid CCl_4 and Na^+ solvation in bulk liquid CCl_4.

The paper is organized as follows. In section 2, we briefly describe the potential model constructed and the computational details of the MD simulations. Properties of the bulk liquid CCl_4 and the solvation properties of the Na^+ ion in liquid CCl_4 are presented in section 3. The conclusion is presented in section 4.

2. METHOD

2.1. Potential Model

Based on the density functional theory (DFT), the electronegativity equalization method (EEM) [22] has recently parameterized and validated for atomic charges calculations by Geerlings, De Proft, Langenaeker, Bultinck, *et al.* [23-25]. Moreover, EEM has been successfully applied to molecular mechanics and molecular dynamics simulation by Chelli, Smirnov, *et al.* [26-29]. Recently, the atom-bond electronegativity equalization method (ABEEM) [30-32] proposed by Yang, has also been fused into molecular mechanics (ABEEM/MM), and applied to the water [15,16], ion/water [17-19], as well as biological interesting systems [20,21]. Now we extend ABEEM/MM to describe the interactions for carbon tetrachloride, CCl_4.

In the following, we consider a flexible body and nine-site fluctuating charge interaction model potential for carbon tetrachloride. CCl_4 has its experimental tetrahedral geometry with a C-Cl bond length of 1.77 Å and a Cl-

C-Cl bond angle of 109.47° [33]. The potential for CCl_4 has the following form,

$$
U_{CCl_4} = \sum_{bonds} k_b (r - r_{eq})^2 + \sum_{angles} k_a (\theta - \theta_{eq})^2 + \sum_i \sum_j 4\varepsilon_{ij} \left[\left(\frac{\sigma_{ij}}{r_{ij}} \right)^{12} - \left(\frac{\sigma_{ij}}{r_{ij}} \right)^6 \right]
$$

$$
+ \sum_i \sum_{j \neq i} \left\{ \left[\frac{1}{2} \sum_a \sum_b \frac{q_{ia} q_{jb}}{r_{ia,jb}} + \frac{1}{2} \sum_{a-b} \sum_{g-h} \frac{q_{i(a-b)} q_{j(g-h)}}{r_{i(a-b),j(g-h)}} + \sum_{g-h} \sum_a \frac{q_{ia} q_{j(g-h)}}{r_{ia,j(g-h)}} \right] \right\}
$$

$$
\tag{1}
$$

where the first and second term stand for the energies of C-Cl bond stretching and Cl-C-Cl angle bending, respectively, accounting for the intramolecular vibration. Here, the bond force constant k_b is 232.4 kcal mol^{-1} Å^{-1}, and the angle force constant k_a is 71.75 kcal mol^{-1}deg^{-1}. r_{eq} and θ_{eq} denote the equilibrium values of the bond length and bond angle, as well as r and θ stand for the actual values of bond lengths and bond angles, respectively. The non-bonded interactions between CCl_4 molecules are represented by a sum of van der Waals (the third term) and electrostatic (the fourth term) interactions, in which the Lennard-Jones and Coulombic expressions are employed to describe them, respectively.

The van der Waals interaction between CCl_4 molecules involves carbon-carbon interaction, carbon-chlorine interaction, and chlorine-chlorine interaction. ε_{ij} and σ_{ij} are the Lennard-Jones parameters, and r_{ij} is the distance between atom si and j. The electrostatic interaction sites of CCl_4 monomer are presented by nine sites, five atoms and four C-Cl bonds. q is the ABEEM charge, the subscripts a and b denote atoms, a-b and g-h denote C-Cl bonds, as well as i and j denote CCl_4 molecules. q_{ia} and $q_{i(a\,b)}$ are the ABEEM charges on atom a and bond a-$b$in CCl_4 molecule i, respectively. The bond charge is located on the point that partitions the bond length according to the ratio of covalent atomic radii of two bonded atoms. $r_{ia,jb}$ is the distance between atom a in CCl_4 molecule i and atom b in CCl_4 molecule j.

The Na$^+$- CCl_4 interaction potential is given by

$$
U_{Na^+ - CCl_4} = \sum_i \left\{ \sum_a 4\varepsilon_{I,ia} \left[\left(\frac{\sigma_{I,ia}}{r_{I,ia}} \right)^{12} - \left(\frac{\sigma_{I,ia}}{r_{I,ia}} \right)^6 \right] + \left[\sum_a \frac{q_I q_{ia}}{r_{I,ia}} + \sum_{a-b} \frac{q_I q_{i(a-b)}}{r_{I,i(a-b)}} \right] \right\}
$$

$$
\tag{2}
$$

Here, $\varepsilon_{I,ia}$ and $\sigma_{I,ia}$ are the Lennard-Jones parameters for the interaction between ion I and atom a in CCl_4 molecule i. The Coulomb interaction incorporates the electrostatic interaction between ion I and the nine charge sites of CCl_4 molecule i. q_I is the ABEEM charge on ion I. $r_{I,ia}$ and $r_{I,i(a-b)}$ are the distances between ion I and atom a, as well as bond a-b in CCl_4 molecule i, respectively. Note that ion I denotes the sodium ion and atom a can be the carbon atom or the chlorine atom.

Table 1. ABEEM/MM parameters for CCl₄ molecule

Atom type	ABEEM parameters[a]		van der Waals parameters	
	χ^{*b}	$2\eta^*$	σ(Å)	ε(kcal/mol)
C-	3.607	0.529	3.410	0.100
Cl-	2.684	3.500	3.578	0.100
C-Cl	3.929	8.353		

[a]ABEEM parameters are employed to calculate the ABEEM charges, and χ^* and η^* are the valence-state electronegativity and hardness, respectively.
[b]The Pauling electronegativity scaling unit is used.

By fitting the structural properties and binding energies of $(CCl_4)_2$ and $Na^+(CCl_4)_n$ (n=1-4) clusters, we have optimized the potential parameters. The parameters of Na^+ in the present study are taken directly from Reference [19] and CCl_4 parameters are summarized in Table 1. Here we give a brief account of the method and the fit procedure. First, we use ABEEM [30-32] to compute the charges of the system. Then, the ABEEM charges were employed to calculate the Coulomb interactions in Equation (1) and (2). Finally, the total potential energy for the system can be obtained. When there is a change of bond length as well as angle, and relative position of CCl_4 molecules or between the Na^+ cation and CCl_4 molecules, we recalculate the charges by ABEEM, and then recalculate the total potential energy of the system.

2.2. Simulation Details

The molecular dynamics simulations include one small system with the cubic box of size 23.45 Å and one big system with the cubic box of size 35.18

Å for liquid CCl_4 and Na^+ ion in bulk liquid CCl_4, respectively. For liquid CCl_4, the small system consists of 80 CCl_4 molecules, and the large system consists of 270 CCl_4 molecules. For Na^+ ion in bulk liquid CCl_4, one system is 1 Na^+ and 79 CCl_4 molecules, and the other is 1 Na^+ and 269 CCl_4 molecules. The simulations were carried out by modified Tinker program at NVT and NPT ensembles with periodic boundary condition and minimum image convention. Temperature was kept constant by Berendsen algorithm [34].

The equations of motion were solved using the velocity Verlet algorithm with a time step of 1.0 fs. Each system was preliminarily equilibrated for 200 ps run and followed by 600 ps of data collection for later analysis. Several points must be mentioned: (1) the force is only acting on the sodium ion and atoms by the repartition of charges from bonds to atoms; (2) the sodium ion and atoms in the system are randomly assigned velocities appropriate for the temperature of the simulation according to Maxwellian distribution, and they are allowed to move according to Newton's equations of motion, as well as the velocities are adjusted intermittently until the system has reached the desired temperature; (3) the cutoff for the nonbonded interactions is 11.5 Å for the small system and 17.5 Å for the big system, and the nonbonded interactions are truncated using the force shifting [35], where the calculated forces and energies are smoothly shifted to zero at the cutoff distance; (4) the charges were recalculated every ps instead of each time step taking account of the expensive computational time, and the equilibration time and production time are long enough to give an equilibrated system and good statistics.

3. RESULTS AND DISCUSSIONS

3.1. Properties of CCl_4 Dimer and Clusters $Na^+(CCl_4)_n$ (n=1-4)

The structures of the investigated CCl_4-CCl_4 and clusters $Na^+(CCl_4)_n$ (n=1-4) are shown in Figure 1. The minimum interaction energy for CCl_4 dimer is about -2.2 kcal/mol with a C---C separation of 4.6 Å calculated by ABEEM/MM force field. Chang *et al.* [14] reported that the binding energy of MP2/aug-cc-pVDZ calculations with counterpoise (CP)-corrected is -3.3 kcal/mol and the C---C equilibrium distance is 4.63 Å, moreover, the polarizable force field interaction energy is -2.9 kcal/mol and the C---C equilibrium distance is 4.6 Å.

From the comparisons, we can see that the ABEEM/MM minimum interaction energy for CCl_4 dimer is a little higher, while the equilibrium distance is in good agreement with the *ab initio* calculation and the polarizable force field result.

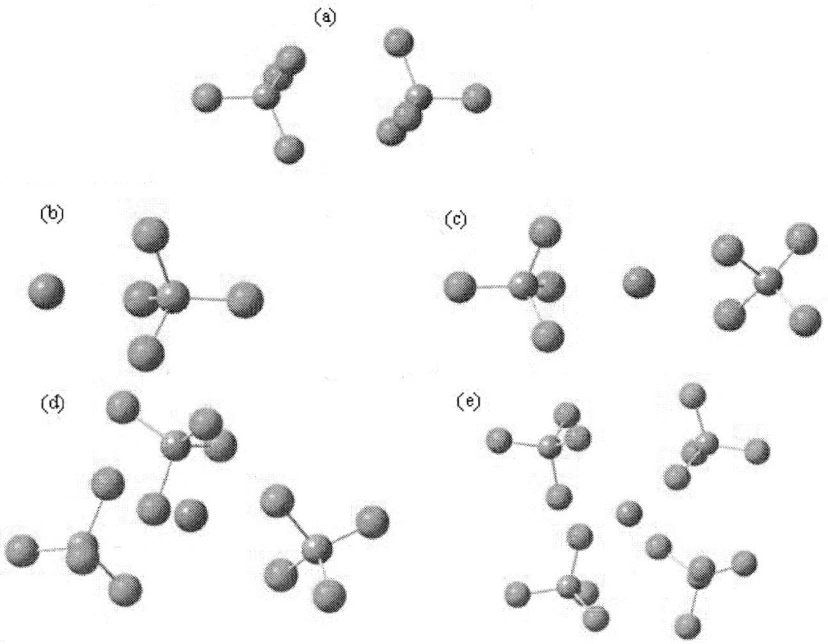

Figure 1. The minimum structures of CCl_4-CCl_4 dimer and clusters $Na^+(CCl_4)_n$ (n=1-4).

The calculated binding energies of $Na^+(CCl_4)_n$ (n=1-4) are -11.30, -21.93, -31.04, and -36.12 kcal/mol, respectively, and they are also in good agreement with the B3LYP/6-31+G(d,p) results (-12.76,-23.59,-31.77,-35.38 kcal/mol, respectively).

3.2. Static Properties of Liquid CCl_4 and Ionic Solution

The calculated average density of liquid CCl_4 is 1.583 g/cm^3 for the small system and 1.581 g/cm^3 for the big system, which are in good agreement with the experimental value 1.584 g/cm^3. In liquid CCl_4, the average C-Cl bond length and Cl-C-Cl angle are 1.771±0.001 Å and 109.399±0.006° for the small

  Xin Li, Li-Dong Gong and Zhong-Zhi Yang

system, respectively, while for the large system they are 1.769±0.001 Å and 109.401±0.004°, respectively. In Na^+ ionic liquid CCl_4, the average C-Cl bond length and Cl-C-Cl bond angle are 1.770±0.001 Å and 109.400±0.005° for the small system, while for the large system they are 1.769±0.001 Å and 109.401±0.004°, respectively. These results are also very close to the experimental value.

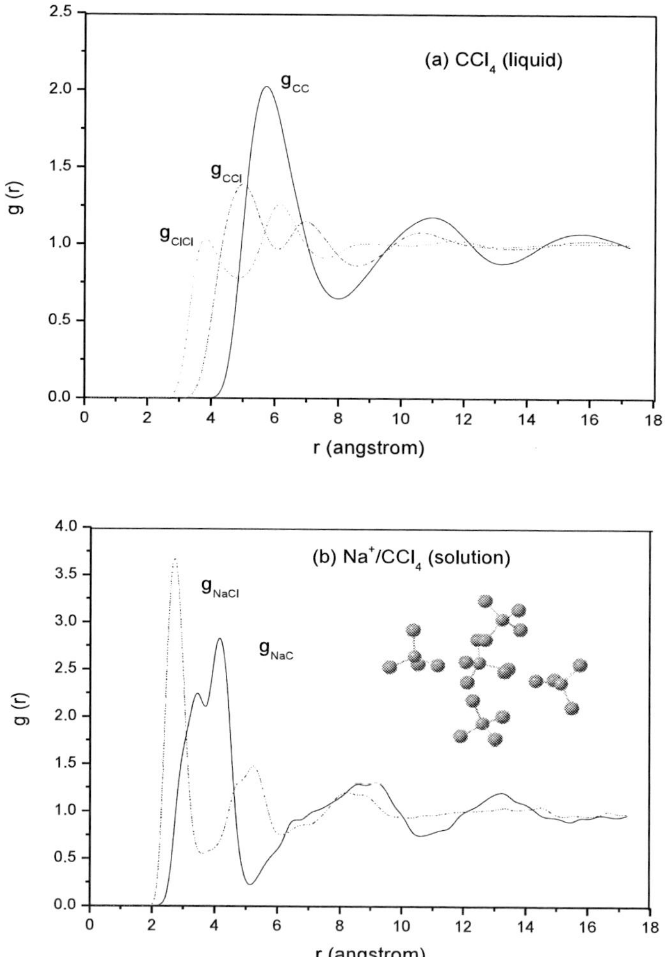

Figure 2. The radial distribution functions $g_{CC}(r)$, $g_{CCl}(r)$, $g_{ClCl}(r)$, $g_{NaC}(r)$, and $g_{NaCl}(r)$.

Therefore, the harmonic potential is good enough to describe the bond stretching and the angle bending. The following simulation results are from the statistics of the big system.

The structure of liquid CCl_4 and a sodium ion in CCl_4 solution can be characterized in terms of the atomic radial distribution functions.

In Figure 2, we display the computed C-C, C-Cl, Cl-Cl, Na^+-C, and Na^+-Cl radial distribution functions. In general, the features of these radial distribution functions, respectively, are consistent with those reported in the literature of different CCl_4 interaction potentials. Integrating over the fist peak of $g_{CC}(r)$ gives about 12 neighboring carbon atoms in the first coordination shell of a central carbon atom. Integrating over the first minimum in $g_{NaC}(r)$ gives about five coordinated CCl_4 molecules in the first solvation shell of Na^+, as compared to the coordination of six water molecules around Na^+ in the aqueous solutions.

The difference comes from the fact that CCl_4 is a much bigger molecule than H_2O, and it is unfavorable to pack as many CCl_4 molecules around Na^+ due to the steric effects. It is also clear from Figure 2 that there is a long-range oscillation located well beyond 15 Å for $g_{CC}(r)$. Compared to that of $g_{CC}(r)$, the correlations of $g_{CCl}(r)$ and $g_{ClCl}(r)$ have a much shorter range.

The first two peak positions of $g_{CCl}(r)$ are 4.95 and 7.00 Å, as well as those of $g_{ClCl}(r)$ are 3.75 and 6.20 Å, in excellent agreement with the experimentally deduced values of 4.9 and 7.2 Å for $g_{CCl}(r)$, and of 3.8 and 6.2 Å for $g_{ClCl}(r)$ [36]. For $g_{NaC}(r)$ and $g_{NaCl}(r)$, there are the characteristic maximum and minimum, which suggests that the CCl_4 molecules form a well-defined solvation shell around Na^+.

In addition, the first and second peaks of $g_{NaCl}(r)$ are in the first shell of Na^+ and the height of the fist peak is 2~3 times higher than that of the second peak, indicating that three chlorine atoms of the CCl_4 molecule face to Na^+, the fourth chlorine atom is in the opposite direction.

In order to examine the local orientational structure between CCl_4 molecules, we have studied the probability distribution of the angle between the intramolecular C-Cl bond and the vector connecting the carbon atom to another carbon atom. Figure 3 depicts the angular distribution functions, P (angle, r), as a function of C---C distance.

It is clear that when the C---C distance is less 5 Å, the interlocking configuration is the most important as indicated by the peaks at 65° and 162° in the angular distribution function. When the C---C distance increases to about the first peak position in the $g_{CC}(r)$, other configurations start to show significant contribution while the interlocking structure remains important.

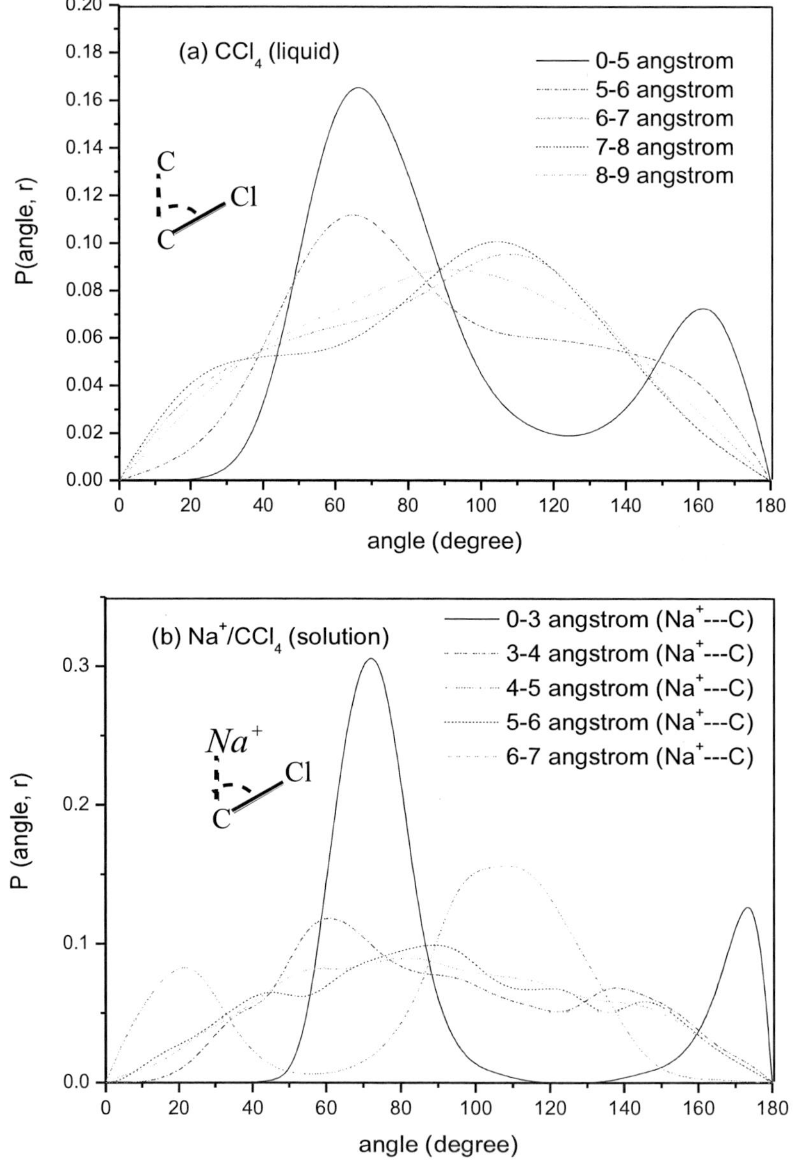

Figure 3. The angular distribution functions, P (angle, r).

As the C---C distance increases, the angular distribution functions approach the sine function that describes a uniform orientation, indicating the vanishing of the orientational correlation. In Na$^+$ ionic solution, the orientation of CCl_4 molecules surrounding Na^+ is examined via the angular distribution function as a function of Na^+---C distance and is also shown in Figure 3.

The angle is defined between the intramolecular C-Cl bond and the vector connecting the Na^+ and the carbon atom. When the Na^+---C distance is smaller than the first peak position of $g_{NaC}(r)$, the computed angular distribution functions show dramatic deviations from the sine curve that describes a uniform distribution. Two peaks centered at 70° and 175° are observed for a Na^+---C distance of less than 3 Å, while the probability of finding CCl_4 in another orientation is almost vanished. Integrating over the area under these two peaks gives a ratio of magnitude of roughly three to one. This implies that each CCl_4 molecule has three of its chlorine atoms facing Na^+, while the fourth chlorine atom locates on the opposite site of the central carbon atom away from Na^+. The above result suggests that Na^+ induces a strong local orientational order in liquid CCl_4. As the Na^+---C distance becomes larger, the angular distribution curve approaches a normal sine function, indicating that the orientational order has vanished.

In addition, our fluctuating charge model can give reasonable ABEEM charge distribution. For liquid CCl_4, the average charges (a.u.) of carbon atom, chlorine atom, and C-Cl bond are -0.010 ± 0.003, 0.156 ± 0.010, and -0.132 ± 0.002, respectively. The average charges of ionic solutions are similar to those of liquid CCl_4. The fluctuating charges are corresponding to the changing electrostaticfield; therefore the present model can describe the Coulomb interaction precisely.

3.3. Dynamical Properties of Liquid CCl_4 and Ionic Solution

We can calculate the diffusion coefficient of liquid CCl_4 by monitoring the mean square displacement (MSD) and by exploiting the well-known relation

$$D = \frac{1}{6}\lim_{t \to \infty}\frac{d}{dt}\left\langle \left| R(t) - R(0) \right| \right\rangle \tag{3}$$

where $R(t)$ and $R(0)$ denote the coordinate of CCl_4 molecular mass center at time t and time 0, respectively. The computed diffusion constant of CCl_4 is $(1.50\pm0.01)\times10^{-5}$ cm^2/s, in agreement with the experimental result of 1.3×10^{-5} cm^2/s [4]. In the Na^+ ionic solution, the calculated diffusion constant of CCl_4 is $(1.16\pm0.02)\times10^{-5}$ cm^2/s, also in agreement with the experimental value.

In one word, the good agreement between various properties from ABEEM/MM results and the available experimental data demonstrates that the present CCl_4 model potential gives a good description of the CCl_4 interaction and Na^+-CCl_4 interaction in bulk liquid and Na^+ ionic solution.

CONCLUSION

Based on the atom-bond electronegativity equalization method fused into molecular mechanics (ABEEM/MM), the nine-site fluctuating charge and flexible body CCl_4 model and Na^+- CCl_4 interaction potential parameterized by gas-phase dimer CCl_4-CCl_4 and ionic clusters $Na^+(H_2O)_n$ (n=1-4) have been tested in the liquid CCl_4 and ionic solution simulations. Although this investigation is far from being complete, the obtained results display a few important observations. The C-Cl bond length and Cl-C-Cl bond angle are very close to the experimental values, indicating that the harmonic potential is good enough to describe the bond stretching and angle bending. The density of liquid CCl_4 agrees well with the experimental value. Simulations with our model provide a local orientational structure, which is in good agreement with the results obtained from other simulations. Moreover, we reproduce the experimental diffusion coefficient of CCl_4 molecules in liquid CCl_4 and ionic solution.

Overall, the central idea of this potential model is to treat charges as variables responding to their environment in a way similar to the polarization in response to real molecules. Compared with the fixed charge models, the CCl_4 potential model and Na^+-CCl_4 interaction potential, based on ABEEM/MM, improve the prediction of static and dynamical properties of the bulk CCl_4 and ionic solution, and this reflects the suitability of the ABEEM/MM interaction potential model.

ACKNOWLEDGMENTS

The authors greatly thank Professor Jay William Ponder for providing the Tinker program. This work is supported by the grant (No. 20633050 and 20703022) from the National Natural Science Foundation of China.

REFERENCES

[1]   Caldwell J., Dang L. X., Kollman P. A., Implementation of Nonadditive Intermolecular Potentials by Use of Molecular Dynamics: Development of a Water-Water Potential and Water-Ion Cluster Interactions, J. Am. Chem. Soc., 1990, 112, 9144-9147.

[2]   Allen M. P., Tildesley D. J., Computer Simulation of Liquids, Oxford University Press, Oxford, 1987.

[3]   King G., Warshel A., Investigation of the free energy functions for electron transferreactions, J. Chem. Phys., 1990, 93, 8682-8692.

[4]   Chahid A., Bermejo F. J., Carcia-Hernandez M., Martinez J. M., Single-particle dynamics of liquid CCl4: a comparison of molecular dynamics and neutron quasi-elastic scattering results, J. Phys. Condens. Matter, 1992, 4, 1213-1232.

[5]   Ewool K. M., Strauss H. L., Far Infrared spectra and octopole moments of the tetrahalo carbons, J. Chem. Phys., 1973, 58, 5835-5836.

[6]   Majer V., Svab L., Svibida V., Enthalpies of vaporization and cohesive energies for a group of chlorinated hydrocarbons, J. Chem. Thermodyn., 1980, 12, 843-847.

[7]   Garcia-Hernandez M., Martinez J. M., Bermejo F. J., Chahid A., Enciso E., Collective dynamics of liquid carbon tetrachloride studied by inelastic neutron scattering and computer simulation, J. Chem. Phys., 1992, 96, 8477-8484.

[8]   Bermejo F. J., Enciso E., Alonso J., Garcia N., Howells W. S., How well do we know the structure of simple molecular liquids? CCl4 revisited, Mol. Phys., 1988, 64, 1169-1184.

[9]   Narten A. H., Danford M. D., Levy H. A., Structure and intermolecular potential of liquid carbon tetrachloride derived from x-ray diffraction data, J. Chem. Phys., 1967, 46, 4875-4880.

[10] Lowden L. J., Chandler D., Theory of Intermolecular pair correlations for molecular liquids. Applications to the liquids-carbon tetrachloride, carbon disulfide, carbon diselenide, and benzene, *J. Chem. Phys.*, 1974, 61, 5228-5241.

[11] McDonald I. R., Bounds D. G., Klein M. L., Molecular dynamics calculations for the liquid and cubic plastic crystal phases of carbon tetrachloride, *Mol. Phys.*, 1982, 45, 521-542.

[12] DeBolt S. E., Kollman P. A., A theoretical examination of solvatochromism and solute-solventstructuring in simple alkyl carbonyl

compounds. Simulations using statistical mechanical-free energy perturbation methods, *J. Am. Chem. Soc.*, 1990, 112, 7515-7524.

[13] Adan F. S., Banon A., Santamaria J., Monte Carlocalculations for liquid carbon tetrachloride, *Chem. Phys. Lett.,* 1984, 107, 475-480.

[14] Chang T. M., Peterson K. A., Dang L. X., Molecular dynamics simulations of liquid, interface, and ionicsolvation of polar-izable carbon tetrachloride, *J. Chem. Phys.,* 1995, 103, 7502-7513.

[15] Yang Z. Z., Wu Y., Zhao D. X., Atom-bond electronegativity equalization method fused into molecular mechanics. I. A seven-site fluctuating charge and flexible body water potential function for water clusters, *J. Chem. Phys.*, 2004, 120, 2541-2557.

[16] Wu Y., Yang Z. Z., Atom-Bond Electronegativity Equalization Method Fused into Molecular Mechanics. II. A Seven-Site Fluctuating Charge and Flexible Body Water Potential Function for Liquid Water, *J. Phys. Chem. A,* 2004, 108, 7563-7576.

[17] Li X., Yang Z. Z., Hydration of Li+-ion in atom-bond electronegativity equalization method–7Pwater: A molecular dynamics simulation study, *J. Chem. Phys.*, 2005, 122, 084514.

[18] Li X., Yang Z. Z., Study of Lithium Cation in Water Clusters: Based on Atom-Bond Electronegativity Equalization Method Fused into Molecular Mechanics, *J. Phys. Chem.* A, 2005, 109, 4102-4111.

[19] Yang Z. Z., Li X., Ion Solvation in Water from Molecular Dynamics Simulation with the ABEEM/MM Force Field, *J. Phys. Chem. A,* 2005, 109, 3517-3520.

[20] Yang Z. Z., Zhang Q., Study of Peptide Conformation in Terms of ABEEM/MM method, *J. Comput. Chem.*, 2006, 27, 1-10.

[21] Yang Z. Z., Qian P., A study of *N*-methylacetamide in water clusters: Based on atom-bond electronegativity equalization method fused into molecular mechanics, *J. Chem. Phys.*, 2006, 125, 064311.

[22] Mortier W.J., Van Genechten K., Gasteiger J., Electro-negativity Equalization: Application and Parametri-zation, *J. Am. Chem. Soc.* 1985,107:829-835.

[23] Geerlings P., De Proft F., Langenaeker W., Conceptual Density Functional Theory, *Chem. Rev.*, 2003, 103, 1793-1874.

[24] Bultinck P., Langenaeker W., Lahorte P., De Proft F., Geerlings P., van Alsenoy C., Tollenaere J. P., The Electro-negativity Equalization Method II: Applicability of Different Atomic Charge Schemes, *J. Phys. Chem. A,* 2002, 106, 7895-7901.

[25] Bultinck P., Langenaeker W., Lahorte P., De Proft F., Geerlings P., Waroquier M., Tollenaere J. P., The Electro-negativity Equalization Method I: Parame-trization Parameter-ization and Validation for Atomic Charge Calculations, *J. Phys. Chem. A*, 2002, 106, 7887-7894.

[26] Chelli R., Procacci P., A transferable polarizable electrostatic force field for molecular mechanics based on the chemical potential equalization principle, *J. Chem. Phys.*, 2002, 117, 9175-9189.

[27] Chelli R., Ciabatti S., Cardini G., Righini R., Procacci P., Calculation of optical spectra in liquid methanol using molecular dynamics and the chemical potential equalization method, *J. Chem. Phys.*, 1999, 111, 4218-4229.

[28] Smirnov K. S., van de Graaf B., Consistent implementation of the electronegativity equalization method in molecular mechanics and molecular dynamics, *J. Chem. Soc. Faraday Trans.*, 1996, 92, 2469-2474.

[29] Yang Z. Z., Wang C. S., Atom-Bond Electronegativity Equalization Method. 1. Calculation of the Charge Distribution in Large Molecules, *J. Phys. Chem. A*, 1997, 101, 6315-6321.

[30] Wang C. S., Yang Z. Z., Atom–bond electronegativity equali-zation method. II. Lone-pair electron model, *J. Chem. Phys.*, 1999, 110, 6189-6197.

[31] Yang Z. Z., Wang C. S., Atom-bond electronegativity equalization method and its applications based on density functional theory, *J. Theor. Comput. Chem.*, 2003, 2,273-299.

[32] Yang Z. Z., Cui B. Q., Atomic Charge Calculation of Metallo-biomolecules in Terms of the ABEEM Method, *J. Chem. Theory. Comput.*, 2007, 3, 1561-1568.

[33] Bartell L. S., Brockway L. O., Schwendeman R. H., Refined procedure for analysis of electron diffraction data and its application to CCl_4, *J. Chem. Phys.*, 1955, 23, 1854-1859.

[34] Berendsen H. J. C., Postma J. P. M., van Gunsteren W. F., DiNola A., Haak J. R., Molecular dynamics with coupling to an external bath, *J. Chem. Phys.*, 1984, 81, 3684-3690.

[35] Steinbach P. J., Brooks B. R., New Spherical-Cutoff Methods for Long-Range Forces in Macromolecular Simulation, *J. Comput. Chem.*, 1994, 15, 667-683.

[36] Narten A. H., Liquid Carbon Tetrachloride: Atom pair correlation functions from neutron and x-ray diffraction, *J. Chem. Phys.*, 1976, 65, 573-579.

INDEX

Q

R